January 19, 2012   21:32   World Scientific Book - 9in x 6in

LECTURES ON
DISCRETE MATHEMATICS FOR
COMPUTER SCIENCE

Algebra and Discrete Mathematics

ISSN: 1793-5873

The series ADM focuses on recent developments in all branches of algebra and topics closely connected. In particular, it emphasizes combinatorics, set theoretical methods, model theory and interplay between various fields, and their influence on algebra and more general discrete structures. The publications of this series are of special interest to researchers, post-doctorals and graduate students. It is the intention of the editors to support fascinating new activities of research and to spread the new developments to the entire mathematical community.

Algebra and Discrete Mathematics Vol. 3

LECTURES ON
DISCRETE MATHEMATICS FOR
COMPUTER SCIENCE

Bakhadyr Khoussainov
University of Auckland, New Zealand

Nodira Khoussainova
University of Washington, USA

We World Scientific

NEW JERSEY · LONDON · SINGAPORE · BEIJING · SHANGHAI · HONG KONG · TAIPEI · CHENNAI

Published by

World Scientific Publishing Co. Pte. Ltd.

5 Toh Tuck Link, Singapore 596224

USA office: 27 Warren Street, Suite 401-402, Hackensack, NJ 07601

UK office: 57 Shelton Street, Covent Garden, London WC2H 9HE

British Library Cataloguing-in-Publication Data
A catalogue record for this book is available from the British Library.

Algebra and Discrete Mathematics — Vol. 3
LECTURES ON DISCRETE MATHEMATICS FOR COMPUTER SCIENCE

ISBN-13 978-981-4340-50-2
ISBN-10 981-4340-50-2

Printed in Singapore.

We dedicate this book to our parents and grandparents

Preface

About the book: This textbook is a result of many discussions between us, the authors. The first author (Bakh) is a well-established mathematician, and the second author (Nodira) is a young computer science PhD candidate. One is an expert in the field, and the other is a student. The first author favors rigorous definitions and proofs, and prefers formal explanations over informal ones. The second author prefers informal explanations with examples. She questions every definition and proof, and asks for the motivation behind each concept. These two views of the authors are complementary, and the synergy between the two perspectives is seen throughout the book. Along with every definition there is an example, along with every proof there is a discussion, and every chapter comes with real-world programming exercises. The textbook brings together two different viewpoints to create a unified textbook on discrete mathematics designed for students in computer science, software engineering, and mathematics.

Before we started writing this textbook, our conversations about computer science and mathematics often returned to a few fundamental questions. What is a definition? What is a theorem? What is a proof? How do we explain and motivate these concepts to students? Since algorithms lie at the heart of computer science and software engineering, our discussions evolved to become more focused on algorithms and their correctness. The questions arising were now the following. What is an algorithm? What does it mean for an algorithm to be correct? How does one prove the correctness of an algorithm? What should be emphasized? Why do we need to learn correctness proofs?

We soon realized that by putting our discussions into writing, we had started the preparation of a textbook. We designed this textbook for early stage undergraduate computer science, software engineering and mathematics students. For computer science and software engineering students, the book provides the mathematical background needed to reason about algorithms, programs and their properties. For instance, the book contains many examples of correctness proofs for algorithms. For mathematics students, the book presents mathematical ideas by connecting them to fundamental concepts in computer science. For instance, the book exposes close connections between inductive proofs and the concept of iteration used in many programming languages.

As our discussions continued, we often found ourselves discussing the similarities between the process of constructing a proof and the process of designing and writing a program. The first step in both processes is to obtain a clear understanding of the problem at hand. Furthermore, for both tasks one needs to familiarize oneself with the existing knowledge and utilize it. For programs, this means knowing previously defined classes, methods and existing application programming interfaces. For proofs, this includes knowing previously proven theorems, definitions, and techniques. Along the same lines, both proofs and algorithms are often designed piece-by-piece. In proving a theorem, we often break down the problem into subproblems and try to solve them first, thus constructing lemmas. Similarly, in constructing a program, we break it down into various methods and classes. Yet another similarity is the use of examples to acquire an intuition for the algorithm or proof that one is trying to construct. All these similarities showcase the prevalence of transferable skills between discrete mathematics and computer science, thus highlighting the importance of a textbook like this one.

Book highlights: We cover a wide variety of topics in discrete mathematics. However, we would like to emphasize some particular aspects of the textbook. The first is induction. The book showcases many uses of induction, including inductive definitions of objects (such as trees, formulas of propositional logic, and regular expressions), as well as various types of proofs by induction. We think that induction is an essential method for reasoning about algorithms, programs and the objects defined by them. This is because induction is a mathematical tool that reflects the concept of iteration, which is pervasive throughout computing.

The second feature is the numerous algorithms and their proofs of correctness. Almost every lecture presents and analyzes at least one algorithm. Most theorems are proven through the analysis of algorithms. Furthermore, because most algorithms use iteration, many of the proofs of correctness utilize induction via the loop invariant theorem, thus giving the reader more practice with induction.

The third highlight is the comprehensive coverage of finite automata. Finite automata constitute a simple yet powerful mathematical model of programs. Via finite automata one can discuss state transitions of programs and their representations, consider various types of design problems, talk about simulating one program by another, and study connections between various ways of defining problems. The lectures on automata appear near the end of this book. This organization helps the reader to apply the knowledge of graphs, algorithms, and induction acquired earlier on in order to learn about finite automata and their properties.

The fourth aspect of the textbook is that every lecture is followed by a set of exercises, including programming exercises. These allow the student to better understand the concepts presented in the lecture, and to apply the newly-acquired knowledge to real problems.

Book organization: This book is designed as a textbook for a single semester course in discrete mathematics. It consists of thirty three chapters, each chapter covering a 50-minute lecture on average. We aimed to keep the book compact, and at the same time clear and easy to understand. The reader can easily flip through the book to gain some sense of the material covered. However, we now briefly outline the key topics covered in this book.

Lectures 1 through 5 are introductory. They introduce the concepts of definitions, theorems, and proofs through integer and modulo arithmetic. Key topics include the fundamental theorem of arithmetic and the Euclidian algorithm. Lecture 5 covers the RSA encryption method, and can be included or omitted at the teacher's discretion. Lectures 6 through 9 study graphs and trees, as well as algorithms on these structures. These are introduced early because they are easy to explain, intuitively understandable, and relate to real-world problems. These lectures also prepare the reader for set-theoretic notations. Key topics in these chapters are the path problems, inductive definition of trees and proofs of some of their properties using induction. Lectures 10 through 13 introduce sets and relations. Unlike traditional textbooks in discrete mathematics, we present sets and relations as a useful language to reason about databases. We hope this keeps the readers engaged and interested, since they immediately see the application of sets to something concrete and useful. Lectures 14 through 19 present induction, inductive proofs, recursion, and correctness of algorithms. The key feature is the loop invariant theorem. These lectures prove the correctness of many algorithms, including Prim's minimum spanning tree algorithm and

Djikstra's shortest path algorithm. Lectures 20 and 21 study functions. In addition to the traditional topics such as surjective, injective and bijective functions, there is a discussion of transition functions. This connects functions with the forthcoming lectures on finite automata. Lectures 22 through 24 study propositional logic. In these lectures, we employ induction to reason about the propositional calculus. The key topics include the unique readability theorem, normal forms, logical equivalence, and models. Lectures 25 through 30 introduce finite automata. The lectures put an emphasis on designing automata, determinism and non-determinism, converting non-deterministic automata to deterministic automata, as well as regular expressions. The key topics presented include the determinization theorem, Kleene's theorem, and algorithms for checking various properties of automata. Finally, Lectures 31 through 33 introduce the basics of counting and probability. The topics include counting rules and principles, permutation and combinations, definition of probability, and probability distributions. Each lecture contains many examples.

Acknowledgments: Parts of this textbook have been used in several courses at the University of Auckland, National University of Singapore, and Cornell University. We would like to thank the students of these classes for testing this book, finding mistakes, and helping us better design the structure of the book. We also thank the following people for their support, comments and suggestions: Cristian Calude, Elena Calude, Michael Dinneen, Gillian Dobbie, Aniruddh Gandhi, Alexander Melnikov, Mia Minnes, André Nies, Eamonn O'Brien, Pavel Semukhin, and Mark Wilson. Without their careful reading of the initial drafts of this textbook, the book would not be possible in its current form. We also thank our friend Igor Polivanyi for his beautiful photograph that we used for the book cover. Finally, we thank you, our reader, for taking the time to read this book. We hope that it is educational, enjoyable and helpful for your computing and mathematical endeavors in the years to come.

Bakhadyr Khoussainov and Nodira Khoussainova

Contents

Lecture 1

Definitions, theorems, and proofs

I'm not telling you it is going to be easy.
I'm telling you it's going to be worth it.
Art Williams.

1.1 Definitions

Before writing a program, we must fully understand what the program is supposed to do. A description of the things that we would like our program to do is called the specification of the program. Correctly writing the specifications of a program requires a lot of time and intellectual effort. However, it is a necessary and helpful task. If we write our specification badly, then the corresponding program is usually hard to understand, incorrect or exhibits undesired behavior. We can use many different techniques and languages for preparing program specifications.

One way to describe the specifications is to explain them informally through conversations and documents, omitting many details. Informal specifications are a helpful first step in writing programs because they do not require the programmer to flesh out all the details of the program a priori and thus can be written quickly. That said, using only this method can have negative consequences. It often leads to an inaccurate understanding of the specifications and their different interpretations. Therefore, such an informal approach can result in incorrect programs.

A second method is to write a semi-formal description of program specifications. Usually, these specifications are written in document format with some structure and a little rigor. These types of specifications are more precise than the informal approach. However, semi-formal specifications can also contain subtle inconsistencies and miss details of the program's requirements. Therefore, semi-formal specifications also leave room for different interpretations and again may lead to writing incorrect programs.

A third approach is to use a formal language to describe the specifications. In this approach, the programmer precisely formulates the program's requirements. This not only helps the programmer clearly understanding the requirements, but it also means that the programmer can use one of the many tools for automatically checking if their specification is consistent, and if it can even be realized. Thus, writing the specifications in a formal language with carefully prepared and precise definitions can be extremely valuable in writing correct programs. However, this is an arduous undertaking and is significantly more difficult than the less formal approaches. Therefore, an effective way to write specifications is to utilize a combination of all three methods above, and to switch back and forth between them as needed.

In modern programming, especially in object-oriented programming, writing code often involves defining abstract objects. These objects and their instances are then manipulated by the methods of the code. Thus, for the code to work and be easy to understand, the programmer must give precise definitions of each object and its methods. This can also help in checking whether the code is correct. Both writing specifications and defining abstract objects require careful thought, clear abstractions, and sound reasoning.

Therefore, learning how to construct precise, clear, and unambiguous definitions is a crucial skill for software engineers and computer scientists. In order to develop the skill of writing good programs and being able to give precise definitions, one needs to have a solid mathematics background and be familiar with formal reasoning. These lectures are designed to introduce the reader to this required background.

In these lectures, we provide many examples of definitions that will be used throughout the course. We separate our definitions in the text and the objects we define will be emphasized in bold **like this**. We also emphasize certain words *like this* in order to bring the reader's attention to significant terms in our discussion.

Natural numbers will be one of the fundamental objects that we use throughout these lectures. *Natural numbers*, which you have already encountered in elementary mathematics, are 0, 1, 2, 3, 4, 5, ... In some textbooks, 0 is not considered a natural number. It is not important whether or not 0 is a natural number as long as we are consistent with our conventions and definitions. In these lectures, we will always consider 0 to be a natural number.

We also use integers throughout the lectures. Recall that these are natural numbers together with their negatives. Thus, *integers* are $0, 1, -1, 2, -2, 3, -3, 4, -4, 5, -5, \ldots$.

Obviously, every natural number is an integer but not every integer is a natural number. *Integer variables* are usually denoted by lowercase letters, such as n, m, t, s, etc. For example, if we consider n, then n represents one of the integers. Thus, n can be $0, 1, -17$ or 2^{32}.

We use the following notation. The collection of *all* natural numbers is denoted by $\mathbb{N}$. Thus, 0 is in $\mathbb{N}$, 2 is in $\mathbb{N}$, and so are 17, 2^{32}, 101, etc. The collection of *all* integers is denoted by $\mathbb{Z}$. Thus, 0 is in $\mathbb{Z}$, -2 is in $\mathbb{Z}$, and so are 17, 2^{32}, -101, etc. As we mentioned earlier, every natural number s that belongs to $\mathbb{N}$, also belongs to $\mathbb{Z}$. Therefore, we say that $\mathbb{N}$ *is a subset of* $\mathbb{Z}$, and write it as $\mathbb{N} \subseteq \mathbb{Z}$. The notation $\subseteq$ is a standard notation for other types of abstract objects and collections, and we will introduce this more formally in Lecture 10. Knowing the natural numbers and the integers, we now define rational numbers as follows.

Definition 1.1. A number is a **rational number** if it can be written as
$$\frac{n}{m}$$
where n, m are integers and $m \neq 0$. In this case, n is called the **numerator** and m the **denominator** of the rational number. Sometimes we represent rational numbers as n/m instead of $\frac{n}{m}$.

Every integer is a rational number. Examples of rational numbers are
$$1, \ -5, \ 0, \ \frac{3}{5}, \ -\frac{9}{4}, \ 101, \ 2\frac{3}{1112}, \ \text{etc.}$$
We denote rational numbers by letters r, q, etc. These can also be thought of as rational number variables. The notation for the collection of all rational numbers is $\mathbb{Q}$.

In order to state that m is a natural number we write $m \in \mathbb{N}$ (i.e. m belongs to the collection of natural numbers $\mathbb{N}$). Similarly, we write $m \in \mathbb{Z}$ and $p \in \mathbb{Q}$ to state that m is an integer and p is a rational number, respectively. This *membership* notation is quite standard for other collections of objects, and we will discuss this in Lecture 10.

Now we walk through some more definitions that will be used throughout this course.

Definition 1.2. An **alphabet** is a finite collection of symbols.

An example of an alphabet is the binary alphabet $\{a, b\}$. This alphabet has two symbols a and b. Note how we described this alphabet; we opened the curly braces $\{$, wrote down the symbols of the alphabet a and b separated by a comma, and then closed the braces $\}$. This is called *set notation*. Here are other examples of alphabets: $\{0, 1\}$, $\{0, 1, b\}$, and $\{\star, \square, a\}$. The collection of all the symbols that can be used in Java code is also an example of an alphabet. In writing alphabets the *order* of symbols does not matter. For example, the following are all the same alphabet: $\{a, b, c\}$, $\{a, c, b\}$, $\{b, a, c\}$, $\{b, c, a\}$, $\{c, a, b\}$ and $\{c, b, a\}$. When writing down the symbols of the alphabet there is no need to write the same symbol twice.

We denote alphabets by upper case Greek letters such as Σ (read 'Sigma'), Ω (read 'Omega'), etc. In these lectures, we will mostly use Σ to denote alphabets.

Definition 1.3. A **string** over the alphabet Σ is a finite sequence of symbols from Σ.

For instance, *aa*, *aaab*, *bab*, *babbbaa*, *b* are examples of strings over the binary alphabet $\{a, b\}$. Unlike alphabets, the order of the symbols in a string matters. Additionally, a symbol is allowed to appear in a string more than once. The length of each of the above strings is 2, 4, 3, 7, and 1, respectively. Formally, the *length* of a string is the number of symbols that occur in it. The length of a string can be 0, in which case that string is denoted by λ (read 'lambda'). The string λ is called the **empty string**.

1.2　Theorems and proofs

Once we have defined objects (such as rational numbers, alphabets, and strings), we would like to reason about them and to derive some interesting properties of these objects. These properties are usually stated as theorems. We explain these in more detail and present several examples.

A *theorem* is a precise statement that needs to be proved. Usually, these statements say something about the defined objects. Some of these statements are very easy to understand, and can be obvious properties that we notice when we look at the definition of the object. Some theorems require more careful thought. Here are some examples.

If n is an integer then $n + 0 = n$. This is clearly a statement about the addition of integers. It says that adding 0 to any integer n produces n. We accept this as a true fact.

If n and m are integers then $n+m = m+n$. This is, again, a property of the addition of integers. Proving this fact requires two steps. The first step needs to define the addition operation $+$. The second step uses a form of reasoning known as *proof by induction*. We will explain proofs by induction in forthcoming lectures. For these lectures, however, we accept statements like these as true facts.

If n and m are positive integers with $m \leq n$ then there are exactly $n - m + 1$ numbers i such that $m \leq i \leq n$. This property of the natural numbers requires a bit more thought as one needs to understand where the $+1$ is coming from.

Every natural number $n > 1$ can be written as a product of prime numbers. In order to understand this statement we need to first know the definition of prime numbers. After this, it may help to rewrite this statement as follows to make its logical structure more explicit: *"If n is a natural number greater than 1, then it can be written as a product of prime numbers"*. We will prove this statement in Lecture 3.

All the examples above state some properties of integers. These statements are of the form

$$\text{If } H \text{ then } C.$$

H is called the *hypothesis*, and C is called the *conclusion*. For example, consider the statement *If n and m are positive integers with $m \leq n$, then there are exactly $n - m + 1$ numbers i such that $m \leq i \leq n$.* The hypothesis is

$$n \text{ and } m \text{ are positive integers with } m \leq n,$$

and the conclusion is

$$\text{there are exactly } n - m + 1 \text{ numbers } i \text{ such that } m \leq i \leq n.$$

In the previous section we introduced rational numbers. Here is one property of the addition of rational numbers.

If p and q are rational numbers then $p + q$ is also a rational number.

Let us prove this statement!

In order to prove this statement we first need to clearly identify and understand both the hypothesis and the conclusion of this statement. The hypothesis is *p and q are rational numbers*, and the conclusion is *p + q is also a rational number*. Our proof goes as follows.

We assume that the hypothesis is true, that is, p and q are rational numbers. We need to show that $p + q$ is a rational number. We know from the definition of rational numbers that there must exist integers n, m and s, t such that

$$p = \frac{n}{m} \quad \text{and} \quad q = \frac{s}{t},$$

where $m \neq 0$ and $t \neq 0$. Let us now add these numbers using the rule about addition of rational numbers:

$$p + q = \frac{n}{m} + \frac{s}{t} = \frac{n \cdot t + s \cdot m}{m \cdot t}.$$

The number $n \cdot t + s \cdot m$ is an integer and so is $m \cdot t$. Moreover, $m \cdot t \neq 0$ since $m \neq 0$ and $t \neq 0$. We have written $p + q$ as a fraction of two integers with a non-zero denominator. In other words, $p + q$ satisfies the definition of a rational number. Thus, we have proved that $p + q$ is a rational number.

The above is a good example of a proof. A proof is a sequence of precise, unambiguous statements that follow one after the other using logical rules and precise definitions. This sequence of statements does not involve extra hypotheses, references to unclear statements, elements of randomness like tossing a coin, etc. Proofs should be precise, formal, and logical sequences of statements.

Some statements require nontrivial reasoning to prove that they are true statements. Such statements are singled out as theorems. In order to classify a statement as a theorem, the statement must be correct. Hence its correctness needs to be proved. Just like the proof we completed above, the proof of a theorem must consist of a sequence of precise, unambiguous statements that logically follow one after the other. Theorems are typically statements that describe interesting and sometimes unexpected properties of objects that we define.

Some statements may look like theorems even though they are not theorems. For example, consider the following statement:

If n and m are natural numbers, then n − m is a natural number.

The hypothesis of this statement is *n and m are natural numbers*, and the conclusion is *n − m is a natural number*. The statement says that *whichever* values n and m take on from the collection of natural numbers, it will always be the case that $n-m$ is also a natural number. Let us consider the case when n is 1, and m is 5. Then $n - m = 1 - 5 = -4$. Clearly, in this case $n - m$ is not a natural number. Therefore, this statement is false because it is not true for *all* instantiations of n and m by natural numbers. Hence, the statement is not a theorem. The numbers n and m that falsify the statement (in our example $n = 1$ and $m = 5$) are called *counter-examples* to the statement. Note that there can sometimes be many counter-examples to statements that are not theorems.

1.3 Exercises

(1) State which of the following numbers are natural numbers, integers, and rational numbers: 27, -17, $\frac{3}{9}$, -25, $\frac{10}{3}$, 0.

(2) Consider the binary alphabet $\Sigma = \{a, b\}$. Take a string w over Σ, where $w = w_1 \ldots w_n$. So, each of $w_1, \ldots, w_n$ is either a or b and w has length n. The symbol w_i is called the ith symbol of the string. We define the equality of strings as follows. Two strings w and v *are equal* if they both have the same length, and $w_i = v_i$ for each i. Do the following:

  (a) List all strings (over Σ) of lengths 0, 1, 2, 3, and 4.
  (b) How many strings are there of length 3, 4, and 5?
  (c) Write down a formula that calculates the number of strings of length n.
  (d) Write down a formula that calculates the number of strings of length at most n.

(3) Call a string $a_1 \ldots a_n$ over a binary alphabet $\{a.b\}$ a *monotone string* if no b in this string is followed by a. How many monotone strings are there of length 2,4,6? Write down a formula that computes the number of monotone strings of length m.

(4) Consider the following statements:

  (a) If n^2 is an even number then n is an even number.
  (b) If both x and y are rational numbers then $x-y$ is a rational number.
  (c) If both x and y are rational numbers then x/y is rational number.
  (d) If both x and y are odd natural numbers then the product $x \cdot y$ is an odd number.

(e) If the product $x \cdot y$ of natural numbers x and y is even then both x and y are even.

(f) If x is a rational number but not an integer and y is an integer, then $x + y$ is a rational number but not an integer.

(g) If x and y are integers then $(x + y)/2$ is also an integer.

For each of these statements do the following:

(a) Write down the hypothesis and the conclusion of the statement.

(b) Prove the statement or give counter-examples.

(5) Let n and m be natural numbers.

(a) How many natural numbers are strictly between n and $n + m$?

(b) How many natural numbers are strictly between 2^n and 2^{n+1}?

Programming exercises

(1) Write a program that, given n and alphabet Σ as inputs, outputs all strings of length n over the alphabet Σ.

(2) Write a program that, given natural number n, outputs all monotone strings of length n over the alphabet $\{a, b\}$. See Exercise 3.

Lecture 2

Proof methods

> *Among all human discoveries, the discovery*
> *of the mistake is the most important one.*
> Stanislaw Lec.

2.1   Direct proof method

In order to call a statement a theorem, we need to prove that the statement is true. There are several methods that can be used to prove that a statement is true. Unfortunately, there is no recipe that tells us which method we need to use in each particular case. Moreover, the same statement can often be proved in several ways. However, once definitions and theorems are clearly understood, one usually has some idea of what proof methods are needed to prove the statement. In this section, we explain a few proof methods. We start with the direct proof method.

The direct proof method of a statement of the form

"If H   then C"

uses the following pattern. Assume that the hypothesis H of the statement is true. Using logical steps, derive that the conclusion C of the statement is also true. We have already applied this method in the previous lecture in proving the statement:

If p and q are rational numbers then $p + q$ is also a rational number.

We give another example of a direct proof. For this example, we need a few more definitions.

Definition 2.1. Given integers n and m with $m > 0$, we say that m **divides** n if $n = k \cdot m$ for some integer k. In this case, we also say that m **is a factor of** n and that n is a **multiple** of m.

For example, the integers 2, 3, 5, 9, and 90 are factors of 90. From the definition, it is clear that any natural number $m > 0$ is a factor of 0. We will now prove our first theorem. Though the proof of the theorem is simple, it is important to go through the proof, especially for beginners. In this lecture, most of our proofs will be detailed and simple. However, in the forthcoming lectures the proofs will be more involved and sometimes with less detail.

Theorem 2.1. *If n and m are positive integers and m is a factor of n then $m \leq n$.*

Proof.
We give a direct proof of this theorem. The hypothesis of the theorem is that n *and m are positive integers and m is a factor of n.* The conclusion is m *is less than or equal to n.* To prove this theorem, we assume that the hypothesis is true. Therefore, $n = k \cdot m$ for some integer k. The integer k must be positive as both n and m are positive by the hypothesis. Therefore, $k \geq 1$. Hence $1 \cdot m \leq k \cdot m$. Note that $m = 1 \cdot m$ and $n = k \cdot m$. Therefore, $m \leq n$. We proved the theorem. $\square$

We used the symbol $\square$ after we finished the proof. From now on, throughout the book, the $\square$ symbol will always indicate the following: *"The statement is proved or the rest of the proof is left to the reader"*.

2.2   Proof by cases method

The proof by cases method is useful when the hypothesis can be broken into several cases. Given a theorem of the form

If H then C

the proof by cases method uses the following pattern. Suppose that H can be broken into several cases, say into three cases, and let us call them H_1, H_2 and H_3. To prove the statement, we need to prove that each of these cases implies the conclusion C. Namely, we need to prove: *if H_1 then C, if H_2 then C*, and *if H_3 then C*.

To give an example, once again, we start with a new definition.

Definition 2.2. The **absolute value** of number x is x if $x \geq 0$ and $-x$ if $x < 0$.

We denote the absolute value of x by $|x|$. Thus, for example $|-0.5| = 0.5$ and $|2| = |-2| = 2$. Clearly, the absolute value of any number is non-negative. We prove the following theorem using the proof by cases method.

Theorem 2.2. *For all numbers x and y we have $|x \cdot y| = |x| \cdot |y|$.*

Proof.
The hypothesis of this theorem is x *and y are numbers*. The conclusion is $|x \cdot y| = |x| \cdot |y|$. For the numbers x and y there are four cases:

(1) $x \geq 0$ and $y \geq 0$,
(2) $x \geq 0$ and $y < 0$,
(3) $x < 0$ and $y \geq 0$, and
(4) $x < 0$ and $y < 0$.

To prove the theorem, we need to show that each of these cases implies the conclusion.

Case 1: $x \geq 0$ and $y \geq 0$. In this case $x \cdot y \geq 0$, and therefore

$$|x \cdot y| = x \cdot y = |x| \cdot |y|.$$

Case 2: $x \geq 0$ and $y < 0$. In this case $x \cdot y \leq 0$, and therefore

$$|x \cdot y| = -(x \cdot y) = x \cdot (-y) = |x| \cdot |y|.$$

Case 3: $x < 0$ and $y \geq 0$. In this case $x \cdot y \leq 0$, and therefore

$$|x \cdot y| = -(x \cdot y) = (-x) \cdot y = |x| \cdot |y|.$$

Case 4: $x < 0$ and $y < 0$. In this case $x \cdot y > 0$, and therefore as above we have:

$$|x \cdot y| = x \cdot y = (-x) \cdot (-y) = |x| \cdot |y|.$$

Note that each of these cases implies the conclusion of the theorem. We point out that in all cases we used the definition of the absolute value. In addition, any pair of numbers x and y fall into one of these four cases. □

Thus, in proving a statement by the proof by cases method, it is important that all the cases considered cover all the possibilities for the hypothesis.

2.3   Proof by contradiction method

This method is quite a fascinating and tricky way to prove a statement. Given a theorem of the form

If H then C

the proof by contradiction method uses the following pattern. Assume that H is true but C is false. Then, using a sequence of logical reasoning, derive a contradiction such as $0 = 1$. Another possible contradiction is to derive that H is false. The reason for deriving the contradiction is the fact that C is assumed to be false. Therefore, C can not be false, and thus it must be true.

To give an example of a proof by contradiction, we need to recall *the real number line*. The real number line can be visualized as a straight horizontal line in the plane. In the real number line there is an initial point representing the integer 0. Integers in this line are represented as points such that these points are evenly spaced on the line. We can refer to these points as integer points in the line. All points x in this real number line, including the integer points, represent numbers. We call these numbers *real numbers*. In particular, every rational number, such as $\frac{1}{2}$ and $-\frac{25}{7}$, is a real number. Let us now give the following definition before continuing onto an example of a proof by contradiction.

Definition 2.3. We say that a real number x is **irrational** if it is not a rational number.

This definition raises interesting questions. Does this definition make sense? Are there numbers that are irrational? In other words, are there points in the real line that do not represent rational numbers?

To answer this question, recall that $\sqrt{x}$ is the number y such that $y^2 = x$ and $y \geq 0$. For example, $\sqrt{16} = 4$. Here is a theorem showing that irrational numbers exist. We prove the theorem by giving an example of an irrational number. We use the proof by contradiction method.

Theorem 2.3. *The number $\sqrt{2}$ is irrational.*

Proof.
The hypothesis of this theorem is *we are given $\sqrt{2}$*. The conclusion is *$\sqrt{2}$ is irrational.* We follow the pattern of the proof by contradiction. So, assume that we are given $\sqrt{2}$ and that $\sqrt{2}$ is not irrational. Then, $\sqrt{2}$ must be a rational number. By the definition of rational numbers there must exist natural numbers n and m such that

$$\sqrt{2} = \frac{n}{m}.$$

We can assume that this fraction *is reduced.* If it is not, then reduce the fraction and proceed. This means that n and m have no factors, greater than 1, in common (e.g. $\frac{1}{2}$ is reduced, but $\frac{3}{6}$ is not). From this equation, by squaring both sides, we have

$$2 \cdot m^2 = n^2.$$

This means that n^2 is an even number. Since n^2 is an even number then it must be the case that n is an even number. So, we can write n as $2k$, where k is a positive integer. Putting $2k$ instead of n in the last equation we deduce that

$$m^2 = 2 \cdot k^2.$$

Again, m^2 is an even number, and hence m must also be an even number. We conclude that n and m are even numbers. Therefore both n and m have a common factor 2. But, we had n and m such that n and m have no factors in common since the fraction n/m was selected to be in reduced form. We arrived at a contradiction. The reason for this contradiction is that we assumed that $\sqrt{2}$ is a rational number. Hence, $\sqrt{2}$ must be an irrational number. □

2.4　Proof by construction method

Some statements state that certain objects exist. To prove such statements, one can provide examples of the objects that are claimed to exist. A simple example is the following statement: *For every natural number n there exists a prime number p greater than n.* In this example, to prove the statement one needs to provide a method that, given a positive integer n, builds a prime number greater than n. We will not prove this statement yet since properties of prime numbers are discussed in the next lecture. We, however, give another example.

Theorem 2.4. *For every natural number n there exists a natural number m greater than n such that the remainder of m when divided by 5 is 1.*

Proof.
The hypothesis of the theorem is that *a natural number n is given*, and the conclusion is that *there exists a natural number m greater than n such that the remainder of m when divided by 5 is 1.* To prove the statement we need to provide a method that constructs a desired m. Consider the value $m = 5n + 1$. We claim that m is a desired number. Indeed, it is obvious that $m > n$. When we divide m by 5, it is also clear the result gives the remainder 1. □

All the proof methods explained so far handle statements of the form *if H then C.* The proofs *always* start with the assumption that the hypothesis H is true. Based on this, our proofs then provide a sequence of logical reasoning that deduces C. Students often ask why we do not consider the case when H is false. When H is false we do not need to prove or disprove that C is true. Therefore, the statement *if H then C* becomes *true* by default.

2.5　Statements with *if and only if* connectives

Finally, let us take a look at theorems that use the *if and only if* connective in their statements. Usually, these theorems are written in the form

$$A \text{ if and only if } B.$$

To prove these types of statements one needs to prove two statements. The first statement is *if A then B*, and the second statement is *if B then A*.

Thus, the theorems of the form A *if and only if* B tell us that the statements A and B are equivalent. We give an example, but first we need yet another definition.

Definition 2.4. A positive integer $n > 1$ is **prime** if every factor of n is either 1 or n.

Examples of prime numbers are 2, 3, 11, 29, and 47. Here is a simple theorem about prime numbers stating that prime numbers can not be written as a product of smaller positive integers:

Theorem 2.5. *An integer $n \geq 1$ is prime if and only if it is not the product of two smaller positive integers.*

Proof.
We need to prove the following two statements.

(1) *If an integer $n \geq 1$ is prime then it is not the product of two smaller positive integers.*
(2) *If an integer $n \geq 1$ is not the product of two smaller positive integers, then n is prime.*

We prove the first statement by contradiction. Suppose that n is prime but there exist two positive integers m and k, both smaller than n, such that $n = k \cdot m$. Then $m \neq n$ and $m \neq 1$. Hence, n has a factor that is not equal to n and 1. This contradicts the fact that n is prime. Therefore, n cannot be the product of two smaller positive integers.

We now prove the second statement. Consider any positive factor m of n. Then, there exists a positive integer k such that $n = k \cdot m$. We know that $k \leq n$, $m \leq n$, and k and m both are factors of n. By the hypothesis, it is not allowed that both factors are less than n. Hence, if $k < n$ then $m = n$ and $k = 1$ since $n = k \cdot m$. If $k = n$ then $m = 1$. Hence, every factor m of n is either n or 1. Hence, n is prime. This proves the second statement. These two statements prove the theorem. $\square$

2.6   Exercises

(1) Prove the following statements. In each case, clearly identify the hypothesis and the conclusion of the statement. For each proof, indicate which method(s) of proof you are using.

   (a) The integer -7 is a rational number.
   (b) Every integer is a rational number.
   (c) If p and q are rational numbers then so are $p - q$ and $p \cdot q$.
   (d) The sum of two even numbers is an even number.
   (e) The sum of an even number and an odd number is an odd number.
   (f) If n^2 is an even number then n is also an even number. See the proof of Theorem 2.3.
   (g) The sum of a rational number and an irrational number is an irrational number.
   (h) The product of a non-zero rational number with an irrational number is an irrational number.
   (i) If n is a non-zero integer and m is a factor of n then $m \leq |n|$.
   (j) For all integers x and y we have $|x + y| \leq |x| + |y|$.
   (k) If $(n - m)^k = 0$ then $n = m$, where n, m, k are all integers and $k > 0$.

(2) Prove that $\sqrt{3}$ is an irrational number.
(3) Prove that if n and m are odd integers then $\sqrt{n^2 + m^2}$ is an irrational number.
(4) Prove that there exist irrational numbers a and b such that a^b is rational. (*Hint:* Start with $a = \sqrt{2}$ and $b = \sqrt{2}$).
(5) Prove that between any two rational numbers p and q, where $p < q$, there exists another rational number strictly between them. Conclude that there are infinitely many rational numbers in between p and q. This property is known as the *density* property of the rational numbers.
(6) A **polynomial** over the integers $\mathbb{Z}$ is an expression $p(x)$ of the form

$$a_0 + a_1 x + \ldots + a_n x^n,$$

where the *coefficients* $a_0, \ldots, a_n$ are integers and x is just a symbol. Two polynomials $f(x) = a_0 + a_1 x + \ldots + a_n x^n$ and $g(x) = b_0 + b_1 x + \ldots + b_m x^m$ are *equal* if $n = m$ and $a_0 = b_0, \ldots, a_n = b_n$. We introduce the following two operations for polynomials $f(x) = a_0 + a_1 x + \ldots + a_n x^n$ and $g(x) = b_0 + b_1 x + \ldots + b_m x^m$, where $m \leq n$. Their *sum* denoted by $f + g$ is the polynomial

$$f + g = (a_0 + b_0) + (a_1 + b_1)x + \ldots + (a_m + b_m)x^m + a_{m+1}x^{m+1} + \ldots + a_n x^n.$$

The *product* denoted by $f \cdot g$ is the polynomial

$$c_0 + c_1 x + \ldots + c_{n+m} x^{n+m}$$

such that $c_0 = a_0 \cdot b_0$, $c_1 = a_0 b_1 + a_1 b_0$, $c_2 = a_0 b_2 + a_1 b_1 + a_2 b_0$, $c_3 = a_0 b_3 + a_1 b_2 + a_2 b_1 + a_3 b_0$, etc. Prove that for all polynomials f, g, and h the following equalities are true:

$$f + g = g + f, \ f \cdot g = g \cdot f, \ \text{and} \ h \cdot (f + g) = h \cdot f + h \cdot g.$$

Programming exercises

(1) Write a program that, given a fraction n/m as input, where n, m are integers and $m \neq 0$, outputs its reduced form. See the proof of Theorem 2.3 for the definition of "reduced".

(2) Write a method that, given integers n and m with $m > 0$ as inputs, outputs q and r such that $n = q \cdot m + r$, where $0 \leq r < m$. The number r is called the **remainder** and q the **quotient** when n is divided by m.

Lecture 3

Integers and divisibility

God made integers; all else is the work of man.
Leopold Kronecker.

3.1 Divisibility

We use integers in our everyday lives. We add them, multiply them, and reason about them. Integers are fundamental to both computer science and mathematics. Yet, there are still many unanswered questions about integers. One of the questions is known as the *twin prime conjecture*.

Consider the prime numbers: 2, 3, 5, 7, 11, 13, 17, 19, 23, 29, 31, etc. Call two prime numbers x and y *twins* if $y - x = 2$. Thus, 3 and 5, 5 and 7, 11 and 13, 17 and 19, 29 and 31 are examples of twins. The conjecture states that there are infinitely many twin primes. To this day, it is still unknown whether this conjecture is true.

In this lecture, our goal is twofold. The first goal is to study some of the basic properties of integers. The second is to develop an intuition for manipulating integers and constructing proofs. Let us refresh our memory of some basic properties of the integers.

- For all integers n and m, $n + m = m + n$.
- For every integer n, $n + 0 = n$.
- For all integers n, m, and k, $n + (m + k) = (n + m) + k$.
- For every integer n, $n + (-n) = 0$.
- For every integer n, $n \cdot 1 = n$.
- For all integers n and m, $n \cdot m = m \cdot n$.
- For every integer n, $n \cdot 0 = 0$.
- For all integers n, m, and k, $n \cdot (m \cdot k) = (n \cdot m) \cdot k$.

- For all integers n, m, and k, $n \cdot (m + k) = n \cdot m + n \cdot k$.
- If $n_1 \leq n_2$ and $m_1 \leq m_2$ then $n_1 + m_1 \leq n_2 + m_2$.

Starting with these basic facts, we study other interesting properties of integers.

Recall that an integer $m > 0$ **divides** n (or m is a **factor** of n) if there exists an integer k such that $n = k \cdot m$. Divisibility or equivalently, being a factor, is an important notion for integers. The next theorem collects some basic properties of divisibility.

Theorem 3.1. *Let n, m and k be positive integers. Then:*

(1) n is a factor of n.
(2) If k is a factor of m and m is a factor of n, then k is a factor of n.
(3) If n is a factor of m and m is a factor of n, then $n = m$.

Proof.
The first property is obviously true as $n = 1 \cdot n$. For the second property, since k is a factor of m and m is a factor of n, there must exist integers s and t such that $m = s \cdot k$ and $n = t \cdot m$. Therefore, $n = t \cdot (s \cdot k) = (t \cdot s) \cdot k$. Thus, k is a factor of n. For the last property, we have $m \leq n$ since m is a factor of n. Here we used Theorem 2.1 from the previous lecture. Similarly, since n is a factor of m, we have $n \leq m$. Note that we are exploiting the fact that both n and m are positive. Therefore, $n = m$. $\qquad\square$

We write $m|n$ to indicate that m is a factor of n. Thus, the theorem above can be rewritten as follows. For all positive integers n, m, and k we have:

- $n|n$,
- if $k|m$ and $m|n$ then $k|n$, and finally,
- if $n|m$ and $m|n$ then $n = m$.

The first property is called *reflexivity*, the second *transitivity*, and the third *antisymmetry*.

3.2 Factors

Let us denote the set of all factors of an integer n by $Factors(n)$. For example, $Factors(120)$ consists of the following numbers: 1, 2, 3, 4, 5, 6, 8, 10, 12, 15, 20, 24, 30, 40, 60, and 120. We write this as follows:

$$Factors(120) = \{1, 2, 3, 4, 5, 6, 8, 10, 12, 15, 20, 24, 30, 40, 60, 120\}.$$

So, the set of all factors of integer n is written by first opening with the curly bracket {, followed by listing all the positive factors of n, and finally closing with the curly bracket }. Note that we are using the same set notation that we used for alphabets in the previous lecture.

Our goal is to design a method that, given an integer $n > 0$, outputs all of its factors. Below is an algorithm that we call the $FindFactors(n)$ algorithm. Inputs to the algorithm are integers n.

Algorithm 3.1 $FindFactors(n)$ algorithm

---
(1) Initialize $i = 1$.
(2) If $i > n$ then stop. Otherwise, if i divides n then output i.
(3) Increment i by 1, and then go to line 2.

The algorithm is simple, but it is instructive to analyze the algorithm and show that it is correct. But, what does it mean for an algorithm to be correct? Informally speaking, an algorithm is correct if it satisfies its specifications. For our $FindFactors(n)$ algorithm the specification states that we would like *all* the factors of n and *only* the factors of n to be listed in the output. Thus, the correctness of the $FindFactors(n)$ algorithm requires proofs of the following two statements:

(1) Every output of the algorithm is a factor of n, and
(2) Every factor of n is an output of the algorithm.

Here is one way to prove these statements. First, let us prove that the algorithm will stop and that it only outputs factors of n. On input $n > 0$, line 2 of the algorithm is called exactly $n+1$ times. When this line has been called $(n+1)$ times, the algorithm stops. According to line 2, the algorithm only outputs factors of n. Next, we show that the $FindFactors(n)$ outputs all the factors of n. So, if m is a factor of n, then we know that $m \leq n$. Therefore, when line 2 is called the mth time, $FindFactors(n)$ outputs m. This proves the second statement.

3.3   The fundamental theorem of arithmetic

An important concept in the study of integers is the notion of a prime number. Recall from the previous lecture that an integer p is **prime** if $p > 1$ and there are *exactly* two factors of p: 1 and p. Prime numbers play an important role in integer arithmetic and algorithms. This is because every positive integer $n > 1$ can be written as a product of prime numbers. For example, $63756 = 2 \cdot 2 \cdot 3 \cdot 3 \cdot 7 \cdot 11 \cdot 23$. The decomposition of positive integers into the product of prime numbers tells us that the prime numbers can be viewed as the building blocks of integers. We will now formally state and then prove the following theorem:

Theorem 3.2 (The fundamental theorem of arithmetic).
Every positive integer n greater than 1 can be written as a product of prime numbers.

Proof.
First, note that if n is a prime number, then the theorem is obviously true. This is because the product of the prime number n is simply n itself. Now, to prove the theorem, we analyze the Algorithm 3.2 below. The input for the algorithm is an integer $n > 1$.

Algorithm 3.2 *PrimeFactorization(n)* algorithm

---
(1) Initialize $s = n$.
(2) Find the smallest factor $m > 1$ of s. Output m.
(3) If $s = m$ then stop. Otherwise, set s to be equal to s/m, and go to line 2 above.

For instance, when $n = 31878$, the values of s and m change as follows:

- $s_1 = 31878$ (initialization).
- $s_2 = 15939$ and $m_1 = 2$.
- $s_3 = 5313$ and $m_2 = 3$.
- $s_4 = 1771$ and $m_3 = 3$.
- $s_5 = 253$ and $m_4 = 7$.
- $s_6 = 23$ and $m_5 = 11$.
- $s_7 = 23$ and $m_6 = 23$.

We now study the algorithm. Our analysis will give us a proof of the theorem. Observe that the value of s changes in line 3 after line 2 is executed. Initially, the value of s is n. After that, whenever line 2 is executed, the current value of s is smaller than the previous value of s (except for the last call). Therefore, line 2 is called at most n times. After line 2 is called for the last time, the algorithm outputs the last value of m, and stops as instructed in line 3.

Let t be the number of times that line 2 is executed. Every time line 2 is executed, the algorithm outputs a positive integer. Thus, the algorithm outputs exactly t positive integers. Let us denote them by m_1, m_2, ..., m_t. After each execution of line 2, the value of s changes in line 3 (unless $s = m$, in which case the algorithm stops). Let s_1, s_2, ..., s_t be all these values taken by s. Thus, $s_1 = n$ and

$$s_1 = m_1 \cdot s_2, \quad s_2 = m_2 \cdot s_3, \quad ..., \quad s_{t-1} = m_{t-1} \cdot s_t, \text{ and } s_t = m_t \cdot 1.$$

Note that each m_i is the smallest factor of s_i. We can prove that m_i must be a prime number. Indeed, if m_i were not a prime number, then it would have some factors less than it, and thus would not be the smallest factor of s_i. Since $s_i = m_i \cdot s_{i+1}$ for all $i = 1, 2, ..., t-1$, we have the following equalities:

$$n = m_1 \cdot s_2$$
$$n = m_1 \cdot m_2 \cdot s_3$$
$$...$$
$$n = m_1 \cdot m_2 \cdot ... \cdot m_{t-1} \cdot m_t.$$

Thus, we have written n as a product of prime numbers. □

Now, let us take a positive integer $n > 1$. By the above theorem, we can write n as the product of prime numbers:

$$n = m_1 \cdot m_2 \cdot ... \cdot m_{t-1} \cdot m_t.$$

In this product, each prime number m_i can be repeated one or more times. For example, $1008 = 2 \cdot 2 \cdot 2 \cdot 2 \cdot 3 \cdot 3 \cdot 7$. Therefore, we can write this product as the product of powers of prime numbers. This proves the following theorem.

Theorem 3.3 (Prime number decomposition theorem). *Every positive integer $n > 1$ can be written as*
$$m_1^{i_1} \cdot m_2^{i_2} \cdot ... \cdot m_k^{i_k},$$
where m_1, ..., m_k are distinct prime numbers, and all i_1, ..., i_k are positive integers. □

3.4   The division theorem and greatest common divisors

The main concept of this section is the notion of the *greatest common divisor* of given integers. Before we explain this notion we start with the following two simple observations stated as theorems. The first theorem states that if k divides two integers then k also divides any linear combination of the integers.

Theorem 3.4. *If k divides integers n and m, then k also divides $a \cdot n + b \cdot m$ and $a \cdot n - b \cdot m$, for all integers a and b.*

Proof.
Since k divides n and m there must exist integers s and t such that $n = t \cdot k$ and $m = s \cdot k$. Therefore

$$a \cdot n + b \cdot m = a \cdot t \cdot k + b \cdot s \cdot k = (a \cdot t + b \cdot s) \cdot k \quad \text{and} \quad a \cdot n - b \cdot m = (a \cdot t - b \cdot s) \cdot k.$$

Thus, the integer k is a divisor of both $a \cdot n + b \cdot m$ and $a \cdot n - b \cdot m$.   □

Our second theorem is known as the division theorem. The theorem restates the division method that we learn early in school.

Theorem 3.5 (The division theorem). *Suppose n and m are integers with $m > 0$. Then there exist integers q and r such that $n = q \cdot m + r$ and $0 \leq r < m$.*

Proof.
Before we explain the proof of the theorem, we make the following remarks. The number q stated in the theorem above is called the **quotient** and r is called the **remainder** of n divided by m. The *key point* in the theorem is that the remainder is always smaller than m.

As an example, let $n = 861$ and $m = 8$. Then the division theorem produces $861 = 107 \cdot 8 + 5$. Therefore, the quotient $q = 107$ and the remainder $r = 5$ and $r < m$.

Now we prove the theorem. Say, $n \geq 0$. Consider the sequence

$$0 \cdot m < 1 \cdot m < 2 \cdot m < 3 \cdot m < \ldots$$

of all multiples of m. In this sequence, find $q \cdot m$ which is *the largest* multiple of m less than or equal to n. Thus, for the found integer $q \cdot m$ we have

$q \cdot m \leq n$. Since $q \cdot m$ is the largest multiple of m that is still less than or equal to n, we know that $qm + m > n$. Therefore $0 \leq n - q \cdot m < m$. Set $r = n - q \cdot m$. Then $n = q \cdot m + r$ which is the desired equality. The case when $n < 0$ is proved in a similar way. $\square$

Now we give one of the key definitions in integer arithmetic.

Definition 3.1. Given two positive integers n and m, a **common divisor of n and m** is an integer k that divides both n and m. The **greatest common divisor of n and m**, written $\gcd(n, m)$, is the largest common divisor of n and m.

As an example, consider the numbers 18 and 24. The divisors of 18 are 1, 2, 3, 6, 9, 18. The divisors of 24 are 1, 2, 3, 4, 6, 8, 12, 24. Their common divisors are 1, 2, 3, and 6. Therefore, $gcd(18, 24) = 6$. Clearly, 1 is a common divisor of any two integers n and m.

When n and m are small we can compute $gcd(n, m)$ as follows. Construct two lists; the first is the list of the divisors of n and the second is the list of the divisors of m. Find the largest number that occurs in *both* lists. This largest number is then equal to $gcd(n, m)$. When n and m are large numbers, however, this method is impractical. A practical algorithm for computing $gcd(n, m)$ is known as the Euclidean algorithm and will be discussed in the next lecture.

A notion closely related to the greatest common divisor is the notion of the least common multiple.

Definition 3.2. A positive integer k is a **common multiple** of positive integers n and m if n and m are factors of k. The **least common multiple** of n and m, denoted by $lcm(n, m)$, is the smallest one among all common multiples of n and m.

Clearly, for positive integers n and m, the integer $n \cdot m$ is a common multiple of n and m. However, sometimes the least common multiple is less than $n \cdot m$. For example, for $n = 36$ and $m = 48$, the least common multiple is 288. Exercise 8 tells us how to compute $lcm(n, m)$ if $gcd(n, m)$ is already computed.

3.5   Exercises

(1) Prove or disprove the following statement: *For all integers* n_1, n_2, m_1, m_2, *if* $n_1 \leq n_2$ *and* $m_1 \leq m_2$ *then* $n_1 \cdot m_1 \leq n_2 \cdot m_2$.

(2) Prove that the smallest factor $m > 1$ of any given integer $n > 1$ is prime.

(3) Consider the $FindFactors(n)$ algorithm.

   (a) List all the outputs of the algorithm on numbers 64 and 120.

   (b) For a positive integer n, its length is the number of digits needed to describe it. For example, the lengths of numbers 1, 9, 26, 100 and 102312 are 1, 1, 2, 3, and 6, respectively. If it takes one unit of time to execute each of the lines 2 and 3 in the $FindFactors(n)$ algorithm, how many time units are needed to run the algorithm on inputs of length t? Discuss whether or not this algorithm is efficient in terms of time.

(4) Prove the Division Theorem for the case when $n < 0$.

(5) Let $m = 7$. Apply the Division Theorem for the following values of n: 4, -3, 25, -19. In each case write down the values of q and r stated in the theorem.

(6) Prove that $gcd(n, m)$ always exists.

(7) Find $gcd(n, m)$ for the following pairs of integers: 24 and 32; 44 and 56; and 64 and 96.

(8) Prove that $gcd(n, m) \cdot lcm(n, m) = n \cdot m$ for all positive integers n and m.

(9) Let $p_0, \ldots, p_n$ be the first $n + 1$ prime numbers listed in increasing order.

   (a) Prove that $p_0 \cdot \ldots \cdot p_n + 1$ can not be written as a product of the prime numbers $p_0, \ldots, p_n$ (*Hint: Use the proof by contradiction method*).

   (b) Conclude that there is a prime number distinct from all the first $n + 1$ prime numbers $p_0, \ldots, p_n$.

(10) Prove that if an integer $k > 0$ divides natural numbers n and m, then k divides the remainder obtained by dividing n by m.

(11) Prove that if $gcd(n, m) = 1$ and both n and m divide k, then $n \cdot m$ also divides k.

(12) The fundamental theorem of arithmetic states that every natural number $n > 1$ can be written as a product of primes. Prove that this factorization is *unique* except for the order of the prime factors.

Programming exercises

(1) Implement the $FindFactors(n)$ algorithm.
(2) Design a method that given an integer $n > 1$, writes n as a product of prime numbers.

Lecture 4

Euclidean algorithm and congruence relations

The past is never dead. In fact, it's not even past.

William Faulkner.

4.1 Euclidean algorithm

The Euclidean algorithm was discovered by a Greek mathematician Euclid around 300 BC. The algorithm determines the greatest common divisor of two integers n and m, which we denoted as $gcd(n, m)$. The algorithm is used in programming, integer arithmetic, number theory, and other areas of computing and mathematics. Our goal here is to learn the Euclidean algorithm and prove its correctness. Here is the algorithm, where positive integers n and m are given as inputs. We use the Division Theorem proved in the previous lecture.

Algorithm 4.1 $Euclidean(n, m)$ algorithm

---

(1) If $n = m$ then output n as $gcd(n, m)$ and stop.

(2) If $n > m$ then initialize $a = n$ and $b = m$. Otherwise, initialize $a = m$ and $b = n$.

(3) Apply the Division Theorem to a and b by finding integers q and r such that $a = q \cdot b + r$, where $0 \leq r < b$.

(4) If $r = 0$ then output b as $gcd(n, m)$ and stop. Otherwise, set $a = b$ and $b = r$. Go to line 3.

Our goal is to analyze the algorithm and prove its correctness. Our analysis will consist of three steps. First, we show that the algorithm stops on any given input of positive numbers n and m. Secondly, we show that the algorithm outputs a common divisor of n and m. Finally, we show that the output is the greatest common divisor. This will prove correctness of the algorithm. In our analysis, we always assume that $n \geq m$. The case $m \geq n$ is treated by swapping n and m. First, we start with an example.

Consider the case when the input values of n and m are: $n = 2528$ and $m = 340$. We apply the Euclidean algorithm step by step as presented in the table below. The initial values of a and b are the following: $a_1 = 2528$ and $b_1 = 340$.

Iteration	a	b	Application of the division theorem
1	$a_1 = 2528$	$b_1 = 340$	$2528 = 7 \cdot 340 + 148$
2	$a_2 = 340$	$b_2 = 148$	$340 = 2 \cdot 148 + 44$
3	$a_3 = 148$	$b_3 = 44$	$148 = 3 \cdot 44 + 16$
4	$a_4 = 44$	$b_4 = 16$	$44 = 2 \cdot 16 + 12$
5	$a_5 = 16$	$b_5 = 12$	$16 = 1 \cdot 12 + 4$
6	$a_6 = 12$	$b_6 = 4$	$12 = 3 \cdot 4 + 0$. Here $r = 0$ Output $gcd(2528, 340) = 4$

We begin by proving that the algorithms stops. Lines 1 and 2 of the algorithm are executed exactly once. The initial values of a and b are n and m, respectively (because we assume that $n \geq m$). After each execution of line 3, the algorithm executes line 4. Each time that line 3 is executed, $b < a$ and $b > 0$. Therefore, the division theorem is applicable to a and b. After each execution of line 4, the current value of a is strictly less than the previous value of a. Hence line 3 is executed at most n times. Once line 3 is executed for the last time, line 4 will be executed (also for the last time) and the algorithm will stop.

We want to show that the algorithm outputs the greatest common divisor of n and m. If $n = m$, it is clear that the algorithm outputs the greatest common divisor of n and m. So let us now assume that $n > m$. Let t be the total number of times that line 3 is executed. Let

$$a_1, \ldots, a_t, \qquad b_1, \ldots, b_t, \qquad q_1, \ldots, q_t, \qquad r_1, \ldots, r_t,$$

be the values that the variables a, b, q, and r take in executions of line 3. Note $a_1 = n$ and $b_1 = m$. Between these values we have the following equalities:

$a_1 = q_1 \cdot b_1 + r_1$, where $0 < r_1 < b_1$
$a_2 = q_2 \cdot b_2 + r_2$, where $a_2 = b_1$, $b_2 = r_1$ and $0 < r_2 < b_2$,

$a_3 = q_3 \cdot b_3 + r_3$, where $a_3 = b_2$, $b_3 = r_2$ and $0 < r_3 < b_3$

. . . .

$a_{t-2} = q_{t-2} \cdot b_{t-2} + r_{t-2}$, where $a_{t-2} = b_{t-3}$, $b_{t-2} = r_{t-3}$ and $0 < r_{t-2} < b_{t-2}$

$a_{t-1} = q_{t-1} \cdot b_{t-1} + r_{t-1}$, where $a_{t-1} = b_{t-2}$, $b_{t-1} = r_{t-2}$ and $0 < r_{t-1} < b_{t-1}$

$a_t = q_t \cdot b_t + r_t$, where $a_t = b_{t-1}$, $b_t = r_{t-1}$ and $r_t = 0$.

The output of the algorithm is b_t, so we want to show that $b_t = gcd(n, m)$. Let us first show that b_t is a common divisor of n and m. For this we use the list of equations above. Clearly b_t is a common divisor of a_t and b_t. Applying Theorem 3.4 from the previous lecture, and the fact that $a_t = b_{t-1}$ and $b_t = r_{t-1}$, we obtain that b_t divides a_{t-1}. Hence b_t is a common divisor of a_{t-1} and b_{t-1}. Similarly, b_t is a common divisor of a_{t-2} and b_{t-2}. If we continue this, from the bottom to the top, in the list of equations above, we finally conclude that b_t is a common divisor of a_1 and b_1, and hence of n and m.

Finally, we want to show that b_t is the greatest common divisor of n and m. Let x be any positive integer that divides n and m. We use the equations above to show that x divides b_t. This will clearly show that b_t is the greatest common divisor of n and m. Again, we use Theorem 3.4 from the previous lecture. If x divides a_1 and b_1, then it divides a_2 and b_2. Therefore it must divide a_3 and b_3. Continuing this on we see that it must divide a_t and b_t. Thus, $x \leq b_t$. Therefore, $b_t = gcd(n, m)$.

The Euclidean algorithm and its analysis prove the following theorem:

Theorem 4.1. *The Euclidean algorithm applied to positive integers n and m produces the greatest common divisor of n and m.* $\square$

From the Euclidean algorithm, we derive an interesting fact stated as a corollary of the theorem. This fact is often used in algebra and number theory.

Corollary 4.1. *For positive integers n and m there exist integers s and t such that*

$$s \cdot n + t \cdot m = gcd(n, m).$$

Proof.

The idea of the proof is the following. Perform the Euclidean algorithm on

n and m. Once $gcd(n, m)$ is found, run the algorithm "backwards" to find s and t. More formally, we proceed as follows. We use the equalities in the proof discussed above.

Clearly $gcd(n, m) = r_{t-1}$. Therefore,

$$gcd(n, m) = a_{t-1} - q_{t-1} \cdot b_{t-1}.$$

Now, $a_{t-1} = b_{t-2}$ and $b_{t-1} = r_{t-2} = a_{t-2} - q_{t-2} \cdot b_{t-2}$. Therefore:

$$gcd(n, m) = -q_{t-1} \cdot a_{t-2} + (1 + q_{t-1} \cdot q_{t-2}) \cdot b_{t-2}.$$

In this equation a_{t-2} can be replaced with b_{t-3}, and b_{t-2} can be replaced with r_{t-3} that equals $a_{t-3} - q_{t-3} \cdot b_{t-3}$. Therefore:

$$gcd(n, m) = (1 + q_{t-1}q_{t-2})a_{t-3} + (-q_{t-1} - (1 + q_{t-1}q_{t-2})q_{t-3})b_{t-3}.$$

Thus, by continuing these substitutions and moving from the bottom to the top in the equations (in the proof of the theorem above), we can find integers s and t such that $gcd(n, m) = s \cdot a_1 + t \cdot b_1$.   □

4.2   Modulo p congruence relation

In this section, we introduce the modulo p congruence relation on integers that forms the foundation of modulo arithmetic. Modulo arithmetic was pioneered by Euler and Gauss around the eighteenth and early nineteenth centuries. It is used in number theory, algebra, cryptography, etc. In fact, the fundamental arithmetic operations performed by most computers are modulo arithmetic operations. As another example, we design a secret message passing method in the next lecture that relies on modulo arithmetic.

We start with two examples. For the first example, consider the set $\mathbb{Z}$ of integers. The set $\mathbb{Z}$ of integers can be broken up into two classes of integers: the class of even integers, and the class of odd integers. We write them, respectively, as:

$$\{\ldots, -6, -4, -2, 0, 2, 4, 6, \ldots\} \text{ and } \{\ldots, -5, -3, -1, 1, 3, 5, \ldots\}.$$

Let us denote the first class by $[0]$ and the second class by $[1]$. We know that the sum of even numbers is even, the sum of odd numbers is even, and the sum of an odd and an even number is odd. Also, the product of even numbers is even, of odd numbers is odd, and of an odd and an even number is again even. Using the notation $[0]$ and $[1]$ we can write all these facts as follows:

$$[0] + [0] = [0], \ [0] + [1] = [1], \ [1] + [0] = [1], \text{ and } [1] + [1] = [0],$$

and

$$[0] \cdot [0] = [0], \ [0] \cdot [1] = [0], \ [1] \cdot [0] = [0], \ \text{and} \ [1] \cdot [1] = [1].$$

The arithmetic performed by the equations above is called *modulo 2 arithmetic*. We discuss this more formally shortly.

For the second example, consider the clock arithmetic in which we have the 12-hour clock. The day is divided into 12 hour periods. If it is 8 now, then 9 hours later the time will be 5. Similarly, if it is 11 now then 23 hours later the time will be 10. The *usual addition* would give us $8 + 9 = 17$ and $11 + 23 = 34$, respectively. These are not correct answers because the clock arithmetic *wraps around* every 12 hours. In this example, the modulus is 12, and the clock arithmetic performed is *modulo 12*.

We turn these examples into a mathematical setting. We fix a positive number p, and call it the *modulus*. For example, p can be 12 or 2^{64}.

Definition 4.1. Two integers n and m are **congruent modulo** p, written $n \equiv m \ (mod \ p)$, if p is a factor of $n - m$.

For instance $3 \equiv 7 \ (mod \ 2)$, $11 \equiv 25 \ (mod \ 7)$, and $23 \equiv 79 \ (mod \ 4)$.

Example 4.1. For $p = 2$, integers n and m are congruent modulo 2 if and only if n and m are both even or both odd. For $p = 3$, all integers congruent to 2 modulo 3 are $\ldots, -7, -4, -1, 2, 5, 8, \ldots$. This list consists of exactly those integers that when divided by 3 produce remainder 2. For instance, $-7 = (-3) \cdot 3 + 2$.

The next theorem states some of the simple properties of the modulo p congruence relation.

Theorem 4.2. *Fix a positive integer p (that we refer to as the modulus). The modulo p congruence relation satisfies the following properties:*

(1) Every integer n is congruent to itself modulo p.

(2) If n and m are congruent modulo p then m and n are also congruent modulo p.

(3) If n is congruent to m, and m is congruent to k modulo p then n is congruent to k.

Proof.

The first property is obvious because p is a factor of $n - n = 0$ for every integer n. The second property is also clear because for all integers n and m if $n - m$ is divisible by p then so is $m - n$. For the last part of the theorem we reason as follows. We write $n - k$ as $n - k = (n - m) + (m - k)$. Note that each $n - m$ and $m - k$ is divisible by p. Hence their sum is also divisible by p. $\qquad\qquad\square$

The first property of modulo p congruence relation stated in the theorem is called *reflexivity*. This second property is called *symmetry*. Finally, the third property is called *transitivity*. In our notation all these properties can be restated as follows. For all integers n, m, and k the following are true:

(1) $n \equiv n \ (mod\ p)$,
(2) If $n \equiv m \ (mod\ p)$ then $m \equiv n \ (mod\ p)$, and
(3) If $n \equiv m \ (mod\ p)$ and $m \equiv k \ (mod\ p)$ then $n \equiv k \ (mod\ p)$.

4.3   Congruence classes modulo p

Now our goal is to give an alternative explanation of the modulo p congruence relation. This is given by the next two theorems. Here is the first theorem.

Theorem 4.3. *Integers n and m are congruent modulo p if and only if when divided by p both n and m produce the same remainder.*

Proof.

The theorem uses the connective "if and only if". Therefore, we must prove the theorem in both directions. In one direction, assume that n and m are congruent modulo p. We want to show that when divided by p they both produce the same remainder. By Definition 4.1, we know that there exists an integer k such that $n - m = k \cdot p$. By the Division Theorem, there exist q_1, q_2, r_1 and r_2 such that $n = q_1 \cdot p + r_1$ and $m = q_2 \cdot p + r_2$, where both r_1 and r_2 are non-negative integers less than p. Thus we have the following equalities:

$$n - m = k \cdot p = (q_1 \cdot p + r_1) - (q_2 \cdot p + r_2) = (q_1 - q_2) \cdot p + (r_1 - r_2).$$

Since p is a factor of $n - m$, the integer $r_1 - r_2$ must also have p as a factor. Since $0 \le r_1 < p$ and $0 \le r_2 < p$, the only way $r_1 - r_2$ can be divided by p

is when $r_1 = r_2$.

Now, we prove the reverse direction. Assume that both n and m, when divided by p, produce the same remainder r. Thus, we can write $n = q_1 \cdot p + r$ and $m = q_2 \cdot p + r$. Therefore

$$n - m = (q_1 \cdot p + r) - (q_2 \cdot p + r) = (q_1 - q_2) \cdot p.$$

Hence n and m are congruent modulo p. □

Suppose that we are given p. Applying the Division Theorem to integers n and p, we find q and r such that $n = q \cdot p + r$ with $0 \le r \le p - 1$. By Theorem 4.3, all integers congruent to n modulo p are the ones that have remainder r when divided by p, and they are:

$$\ldots, -3 \cdot p + r, \ -2 \cdot p + r, \ -p + r, \ r, \ p + r, \ 2 \cdot p + r, \ 3 \cdot p + r, \ldots.$$

Obviously, n occurs in this list. We denote this list by $[n]$ and call it the **congruence class** of n (modulo p). Thus, in writing this is expressed as

$$[n] = \{\ldots, -3 \cdot p + r, \ -2 \cdot p + r, \ -p + r, \ r, \ p + r, \ 2 \cdot p + r, \ 3 \cdot p + r, \ldots\}.$$

By Theorem 4.3, if m is congruent to n modulo p then the congruence class $[m]$ produced by m is exactly the same one produced by n. We denote this fact by $[n] = [m]$.

Example 4.2. For $p = 7$, our task is to list all the integers congruent to -16 modulo 7. For this, we apply the theorem above. We divide -16 by 7 and get the equality: $-16 = -3 \cdot 7 + 5$. The remainder is 5. Therefore all integers congruent to -16 modulo 7 are exactly those integers m that when divided by 7 give remainder 5. We can list all those integers and write

$$[-16] = \{\ldots, -30, -23, -16, -9, -2, 5, 12, 19, 26, 33, 40, \ldots\}.$$

The theorem below states some of the fundamental properties of congruence classes:

Theorem 4.4. *Given a modulus p, the congruence classes $[0]$, $[1]$, ..., $[p - 1]$ have the following properties:*

(1) Any integer n is contained in one of the congruence classes. Therefore, there exists an r such that $[n] = [r]$, where $0 \le r \le p - 1$.

(2) No two distinct congruence classes among $[0]$, $[1]$, ..., $[p - 1]$ contain an integer in common.

Proof.

The proof follows from the definitions given above. Indeed, take an integer n, and apply the Division Theorem to n and p to get the equality $n = q \cdot p + r$, where $0 \leq r \leq p - 1$. Therefore, n is contained in $[r]$ the congruence class of r, and $[n] = [r]$ as already observed. The remainder r is unique and therefore n can only occur in $[r]$. $\qquad\square$

The collection of all congruence classes modulo p is denoted by $\mathbb{Z}(p)$. We write this as follows:

$$\mathbb{Z}(p) = \{[0], [1], \ldots, [p - 1]\}.$$

Any integer i contained in the congruence class $[n]$ is called a **representative** of the class. For instance, n is a representative of its own class $[n]$. As an example, let our modulus p be 3. Then we have the following: $[0] = \{\ldots, -6, -3, 0, 3, 6, \ldots, \}$, $[1] = \{\ldots, -5, -2, 1, 4, 7, \ldots, \}$, and $[2] = \{\ldots, -4, -1, 2, 5, \ldots, \}$. Thus, 0, 9, −6 are examples of representatives for the congruence class $[0]$, and 7, −5, 1 are examples of representatives for the congruence class $[1]$.

4.4 Exercises

(1) Find $gcd(n, m)$ for the following pairs of integers using Euclidean algorithm: 24 and 32; 44 and 56; and 64 and 96.
(2) Find s and t such that $gcd(n, m) = s \cdot n + t \cdot m$ for each of the pairs n and m from the previous exercise.
(3) Consider the Euclidean algorithm. For inputs n and m the size of the input is the maximum of the lengths of n and m. Recall that for a positive integer n, its length is the number of digits needed to describe it. For example, the lengths of numbers 1, 9, 26, 100 and 102312 are 1, 1, 2, 3, and 6, respectively.

   If it takes one unit of time to execute lines 3 and 4 of the algorithm, how much time, in terms of time units, is needed to run the algorithm on inputs of length at most t? Discuss whether or not this algorithm is efficient.
(4) Write down all the congruence classes modulo 5 and modulo 7.
(5) Consider modulo 7 congruence relation. Prove the following. If $n \equiv n'$ *(mod* 7*)* and $m \equiv m'$ *(mod* 7*)* then $n + m \equiv n' + m'$ *(mod* 7*)*, $n - m \equiv n' - m'$ *(mod* 7*)* and $n \cdot m \equiv n' \cdot m'$ *(mod* 7*)*.

(6) Prove that if $x \not\equiv 0 \ (mod \ p)$ and $y \not\equiv 0 \ (mod \ p)$, where p is a prime number, then $x \cdot y \not\equiv 0 \ (mod \ p)$. Also, explain why we need to assume that p is a prime number.

Programming exercises

(1) Implement the Euclidean algorithm.
(2) Write a program that on integers n, m and $p > 0$, outputs whether n and m are congruent modulo p.
(3) Write a program that given modulus p and integer k, prints the first k smallest, positive integers from each congruence class modulo p.

Lecture 5

Secret message passing

The personal life of every individual is based on secrecy, and perhaps it is partly for that reason that civilized man is so nervously anxious that personal privacy should be respected.

Anton Chekhov.

5.1 The problem set up

Imagine the following situation. You live in New Zealand and want to secretly communicate with your friend in Russia. To keep the communication secret, you both need to agree on a piece of special knowledge to encode and decode the messages you send. The messages being passed are text. To keep these messages secret, you both need to use a method that makes the messages unreadable to anyone except the two of you. The method is usually called the *cipher* and the process of transforming messages so that they can be secretly shared is called *encryption*. The special knowledge that is needed to encode and decode the messages is called a *key*. To be safe, you should change your key occasionally. This is because using the same key often makes it easier for others to learn your cipher and consequently read your messages.

The goal of this lecture is to explain a well-known encryption method called the RSA encryption algorithm. The encryption method was developed by R. Rivest, A. Shamir and L. Adleman at MIT in 1977. The algorithm has become the most widely accepted method in public-key encryption. For instance, the method is used in data exchange, email security systems, and digital signatures.

Here is an *informal explanation* of how the technique works. To communicate with your friend in Russia you make up two keys: E_1 and D_1. The first key E_1 is for encoding the message and the second key D_1 is for decoding the message. Similarly, your friend in Russia does exactly the same thing. Your friend makes up two keys, E_2 and D_2, for encoding and decoding, respectively. You tell your friend your encoding key E_1 but not your decoding key D_1. Your friend tells you the key E_2 but not D_2. Now the two of you can secretly communicate as follows:

(1) You send your friend a message M_1 by encoding the message using E_2.
(2) Your friend decodes the message using D_2, the key that only your friend knows.
(3) Your friend responds by sending you message M_2 encoded using E_1.
(4) You decode the message by using D_1, the key that only you know.

If you would like to change the keys, then it is not a problem. You both make up new pairs of encoding and decoding keys each. If encoding keys are stolen, then it is not a serious concern. These keys are used for encoding but not for decoding. Very often, encoding keys are called *public keys* and they can be made public, say by publishing the keys in a known location. The idea here is that you are suggesting people to use the public key and send you messages using the key. Since you are the only person who knows how to decode the message, you are the only one who can read the message. The methods D_1 and D_2 are sometimes called *private keys*. It is important that the decoding method be complicated enough so that it is extremely difficult for others to figure out the messages encoded by public keys.

There is one interesting issue here. Your friend would like to be sure that the message obtained is actually from you. Since your friend has made the encoding key E_2 public, anyone can send a message pretending to be you. So a natural question arises as to how your friend can be certain that the message is actually from you. This is known as the *certification problem*.

Here is one way to solve the certification problem. You select a word W that both you and your friend know. This can, for instance, be your name. Pretending that the word you selected is an encoded message, you decode it by using your private key D_1. You are the only person who can do this because you know the key D_1 and others do not. The decoded word will now be a sequence of junk characters. Let us denote this sequence of characters by S. Using your friend's key E_2, you send an encoded message M_1 appended with the sequence S. Your friend decodes the message using

D_2, sees the message M_1 and the sequence S. Using your public key E_1, your friend encodes the sequence S and sees that it is your name. This certifies the message and your friend sees that the message is yours. Once again, it is a good idea to change the word W occasionally, to keep the communication private.

Now we will make mathematical sense of all these secret message passing ideas, keys, and methods. Interestingly, to do this, we need to introduce and study modulo arithmetic based on the notion of modulo p congruence relations introduced in the previous lecture.

5.2   Modulo arithmetic

Consider the set $\mathbb{Z}$ of all integers. Let $p > 1$ be a positive integer, our modulus. Consider the set $\mathbb{Z}(p)$ which we defined in the previous lecture:

$$\mathbb{Z}(p) = \{[0], \ [1], \ \ldots, \ [p-1]\}.$$

Our goal is to introduce the addition and multiplication operation in $\mathbb{Z}(p)$. How would we add and multiply two congruence classes $[n]$ and $[m]$? The intuition is that to add the congruence classes $[n]$ and $[m]$ we first perform the addition operation $n+m$, and then declare the congruence class $[n+m]$ to be the sum $[n] + [m]$. Similarly, to multiply $[n]$ and $[m]$ we first take the product $n \cdot m$, and then declare the congruence class $[n \cdot m]$ to be the product $[n] \cdot [m]$.

We would like these definitions for addition and multiplication *not to be dependent* on the representatives of the congruence classes. In other words, if n' is congruent to n modulo p and m' is congruent to m modulo p, then we would like that $n' + m'$ to be congruent to $n + m$ modulo p. Similarly, we would like that if n' is congruent to n modulo p and m' is congruent to m modulo p, then $n' \cdot m'$ is congruent to $n \cdot m$ modulo p.

The next theorem ensures exactly this behavior and tells us that in order to add and multiply two congruence classes it suffices to add and multiply their representatives.

Theorem 5.1. *Assume that n, m and n', m' are integers such that $n \equiv n' \ (mod \ p)$ and $m \equiv m' \ (mod \ p)$. Then $n + m \ \equiv \ n' + m' \ (mod \ p)$ and $n \cdot m \equiv n' \cdot m' \ (mod \ p)$.*

Proof.

Since n and n' are congruent modulo p there exist integers $q \geq 0$ and $q' \geq 0$ such that $n = q \cdot p + r$ and $n' = q' \cdot p + r$. Similarly, there exist integers $t \geq 0$ and $t' \geq 0$ such that $m = t \cdot p + r_1$ and $m' = t' \cdot p + r_1$. To show that $n + m$ is congruent to $n' + m'$ modulo p, we need to explain why $(n + m) - (n' + m')$ is divisible by p. This is derived from the following calculations:

$$n + m - (n' + m') = [(q \cdot p + r) + (t \cdot p + r_1)] - [(q' \cdot p + r) + (t' \cdot p + r_1)] = (q + t - q' - t') \cdot p.$$

The proof of the theorem for the product operation is left as Exercise 1. $\square$

From this theorem we conclude that we can introduce the addition operation between the congruence classes modulo p. Indeed, let $[n]$ and $[m]$ be congruences classes modulo p. We define $[n] + [m]$ to be $[n + m]$. The theorem above says that this addition is correctly defined because $[n + m]$ does not depend on representatives of the congruence classes of $[n]$ and $[m]$.

By the theorem above, in $\mathbb{Z}(p)$, we can explicitly write down the rule for addition of congruence classes. Indeed, for $[r_1]$ and $[r_2]$ in $\mathbb{Z}(p)$, with $0 \leq r_1 \leq p - 1$ and $0 \leq r_2 \leq p - 1$, we define:

$$[r_1] + [r_2] = \begin{cases} [r_1 + r_2] & \text{if } r_1 + r_2 < p, \\ [r_1 + r_2 - p] & \text{otherwise.} \end{cases}$$

Note that $0 \leq r_1 + r_2 - p < p$ and both $r_1 + r_2$ and $r_1 + r_2 - p$ are congruent modulo p. As an example, let us add 25 and 41 modulo 7. To do this we divide both 25 and 41 by 7. We have $25 = 3 \cdot 7 + 4$ and $41 = 5 \cdot 7 + 6$. Therefore $25 + 41$ is congruent to $4 + 6$ modulo 7. In turn, $4 + 6$ is congruent to 3 modulo 7, and therefore $25 + 41$ is congruent to 3 modulo 7. Thus, we have the following equalities: $[25] + [41] \equiv [4] + [6] \equiv [3] \ (mod\ 7)$.

As an example, the table below is the addition table for $\mathbb{Z}(4)$:

+	[0]	[1]	[2]	[3]
[0]	[0]	[1]	[2]	[3]
[1]	[1]	[2]	[3]	[0]
[2]	[2]	[3]	[0]	[1]
[3]	[3]	[0]	[1]	[2]

One interesting property of the multiplication on congruence classes is the following. Assume that our modulus p is a prime number. Take two numbers x and y such that $x \not\equiv 0 \ (mod\ p)$ and $y \not\equiv 0 \ (mod\ p)$. Then it must be the case that $x \cdot y \not\equiv 0 \ (mod\ p)$. Indeed, if $x \cdot y \equiv 0 \ (mod\ p)$ then $x \cdot y$ is divisible by p. Since p is prime, p must divide either x or y. However, this impossible by the assumption on x and y.

5.3 Three cute theorems

Relatively prime numbers play important role in modulo arithmetic. We will use this concept in our description of the RSA encryption method.

Definition 5.1. We say that two positive integers n and m are **relatively prime** if the greatest common divisor of n and m is 1.

For instance, 12 and 25 are relatively prime. Clearly, any two prime numbers p and q, with $p \neq q$, are also relatively prime.

Here is one simple but important property of relatively prime numbers n and m. Assume that both n and m divide a number s. Hence all divisors of n must divide s. Similarly, all divisors of m must divide s. Since n and m are relatively prime, these two numbers have no divisors, greater than 1, in common. Therefore $n \cdot m$ must also divide s.

Let p be a prime number. We select a positive integer a such that p and a are relatively prime. Consider now the following sequence:

$$a, 2a, 3a, \ldots, (p-1)a.$$

We claim that no two numbers in this sequence are congruent modulo p. Assume that there are two of them, say na and ma, with $n > m$, are such that $na \equiv ma \ (mod \ p)$. Then we have $na - ma = (n-m)a = kp$ for some integer $k > 0$. Since a and p are relatively prime, it must be the case that p divides $n - m$. But $n < p$ and $m < p$. Hence, $n - m$ cannot be divided by p. This is a contradiction. Therefore, all the numbers $a, 2a, 3a, \ldots,$ $(p-1)a$ are pairwise distinct modulo p. This implies that the sequence

$$a, 2a, 3a, \ldots, (p-1)a$$

is simply a rearrangement of the sequence 1, 2, ..., $p-1$ modulo p. Therefore, we have the following equalities all taken modulo p:

$$1 \cdot 2 \cdot \ldots \cdot (p-1) \equiv a \cdot 2a \cdot \ldots (p-1) \cdot a \equiv a^{p-1} \cdot (1 \cdot 2 \cdot \ldots \cdot (p-1)) \ (mod \ p).$$

Hence, some reasoning shows that $a^{p-1} \equiv 1 \ (mod \ p)$ (See Exercise 4).

We put all these observations together as Fermat's Little Theorem:

Theorem 5.2 (Fermat's Little Theorem). *If p is prime and a is a positive integer such that a and p are relatively prime, then $a^{p-1} \equiv 1 \ (mod \ p)$.*

Let n be a positive integer. We count all the positive integers less than n that are relatively prime to n and denote this number by $\phi(n)$. For instance, here are a few values of ϕ: $\phi(2) = 1$, $\phi(3) = 2$, $\phi(5) = 4$, $\phi(6) = 2$, $\phi(8) = 4$, $\phi(10) = 4$. This is known as the *Euler function*. A very simple case to count ϕ is on prime numbers. For prime number p, we have $\phi(p) = p - 1$.

Theorem 5.3 (Euler's Theorem). *If a and m are relatively prime then*

$$a^{\phi(m)} \equiv 1 \ (mod \ m).$$

Proof.
When m is prime (hence $\phi(m) = m - 1$) then the theorem states that $a^{m-1} \equiv 1 \ mod(m)$. Hence, this theorem extends Fermat's Little Theorem. The proof of this theorem is similar to the proof of Fermat's Little Theorem. Let us denote $\phi(m)$ by n, so $n = \phi(m)$. We list all numbers that are relatively prime to and smaller than m: $a_1, a_2, \ldots, a_n$. Consider now the following sequence of numbers modulo m:

$$(\star) \qquad a_1 \cdot a, \ a_2 \cdot a, \ a_3 \cdot a, \ \ldots, \ a_n \cdot a.$$

No two numbers in this sequence are congruent modulo p. This can be shown in a way similar to what we did in the proof of Fermat's Little Theorem. Therefore, the sequence $(\star)$ above is simply a re-arrangement of the sequence $a_1, a_2, \ldots, a_n$ modulo m. Thus we have the following equalities, all modulo m:

$$(a_1 \cdot a_2 \cdot \ldots \cdot a_n) \equiv (a_1 \cdot a) \cdot (a_2 \cdot a) \cdot \ldots \cdot (a_n \cdot a) \equiv a^n \cdot (a_1 \cdot a_2 \cdot \ldots \cdot a_n) \ (mod \ m).$$

Hence, $a^{\phi(n)} = 1 \ (mod \ m)$. $\qquad\qquad\qquad\qquad\qquad\qquad\qquad\qquad\qquad\square$

The final theorem needed for the RSA encryption is the following:

Theorem 5.4 (Chinese Remainder Theorem). *Let n and m be two relatively prime numbers. If $a \equiv b \ (mod \ n)$ and $a \equiv b \ (mod \ m)$ then $a \equiv b \ (mod \ n \cdot m)$.*

Proof.
By the hypothesis of the theorem both n and m divide $a - b$. Since, both n and m are relatively prime, as we noted at the start of this section, $a - b$ is divisible $n \cdot m$. Thus, it is clear that $a \equiv b \ (mod \ n \cdot m)$. $\qquad\quad\square$

5.4 Description of the RSA encryption

In this section, we describe the RSA encryption method of encoding and decoding messages between you and your friend. The method consists of six steps, which we outline below. In describing each step, we discuss the reasoning behind it.

Step 1: Select two prime numbers, call them p and q. In practice, these numbers should be *large*. Multiply these numbers, and get the number $N = p \cdot q$.

Comment. The numbers p and q together form your *private key* that no one else knows.

Step 2: Select another number e that is relatively prime to $(p-1) \cdot (q-1)$. Typically, this number can be small.

Comment. The two numbers N and e form your *public key*. Tell these two numbers to your friend in Russia (or anyone else you want).

Step 3: Suppose your friend wants to send a message M. We assume that M is a number strictly less than $p \cdot q$. Otherwise, M can be broken into smaller pieces and sent piece by piece.

Comment. One can view M as a number since there are many ways to encode text into numbers.

Step 4: Your friend calculates M^e modulo N. Since the calculation is modulo N, the encoded message $C \equiv M^e \pmod{N}$ is a number less than N.

Comment. This number $C \equiv M^e \pmod{N}$ is the encoded message sent to you from Russia. Note that your friend has used your public key N and e.

Step 5: Your job is now to decode message C using your private key: p and q. To do this you find a number d such that

$$e \cdot d \equiv 1 \pmod{(p-1) \cdot (q-1)}.$$

Comment. Here we explain the significance of d. Note that $e \cdot d$ can be expressed as $e \cdot d = 1 + k(p-1)(q-1)$ for some integer $k \geq 0$. We now calculate C^d and obtain the following equalities:

$$C^d = M^{ed} = M^{1+k(p-1)(q-1)} = M \cdot M^{k(p-1)(q-1)} = M \cdot \left(M^{(p-1)(q-1)}\right)^k.$$

Now there are two cases:

Case 1. M is relatively prime with p. By Fermat's Little Theorem we have

$$C^d = M \cdot (M^{(p-1)(q-1)})^k = M \cdot ((M^{p-1})^{q-1})^k \equiv M \cdot (1^{q-1})^k \equiv M \ (mod \ p).$$

Case 2. M is not relatively prime with p. In this case we have

$$C^d = M \cdot (M^{(p-1)(q-1)})^k \equiv 0 \equiv M \ (mod \ p).$$

In a similar way, observe that $C^d \equiv M \ (mod \ q)$. Thus, we have $C^d \equiv M \ (mod \ p)$ and $C^d \equiv M \ (mod \ q)$ and p and q are relatively prime. Apply Chinese remainder theorem to p, q, C^d and M, and get the following equality:

$$C^d \equiv M \ (mod \ N).$$

Step 6: Finally, you compute C^d modulo N. This gives you the message M sent by your friend. You have now decoded your friend's message.

Step 5 requires you to find the integer d such that $e \cdot d \equiv 1 \ (mod \ (p-1) \cdot (q-1))$. In order to find such d we use **Euler's Theorem** above. By the theorem, since e is relatively prime to $(p-1) \cdot (q-1)$, we have the following equality:

$$e^{\phi(e)} \equiv 1 \ (mod \ (p-1) \cdot (q-1)).$$

Hence, we select $d = e^{\phi(e)-1}$. In this case, it is clear that $e \cdot d \equiv 1 \ (mod \ (p-1) \cdot (q-1))$.

We now give an example of an encryption. In our example we selected relatively small prime numbers. Typically it is always better to select much larger numbers, say with around 100 or more digits.

Step 1: Let your private key be $p = 7933$ and $q = 6469$. Then $N = p \cdot q = 51318577$.

Step 2: Consider $(p-1) \cdot (q-1) = 51304176$. Select number e that is relatively prime with 51304176. Set e to be 17. Thus, your public key is $N = 51318577$ and $e = 17$.

Step 3: Assume that your friend wants to send a message M, and say the message M is 941423. Clearly $M < N$.

Step 4: Now your friend encodes the message using the public key N and e. The encoded message is $M^e \ (mod \ N)$. So we calculate M^e modulo N:

$$941423^{17} \equiv 11254751 \ (mod \ 51318577).$$

Step 5: Thus, you receive the encoded message $C = 11254751$. You find a number d such that $e \cdot d \equiv 1 \ (mod(p-1) \cdot (q-1))$. This d must be such that

$$17 \cdot d \equiv 1 \ (mod \ 51304176).$$

In our case $d = 21125249$. To check this, we compute:

$$17 \cdot 21125249 - 1 = 359129232 = 7 \cdot 51304176.$$

Thus, it is indeed the case that $e \cdot d \equiv 1 \ (mod(p-1) \cdot (q-1))$.

Step 6: To decode the message you calculate

$$C^d \equiv 11254751^{21125249} \ (mod \ 51318577).$$

This will give the message $M = 941423$ which is the message sent to you by your friend.

5.5 Exercises

(1) Let n, m and n', m' be integers such that n is congruent to n' modulo p and m is congruent to m' modulo p. Prove that $n \cdot m$ is congruent to $n' \cdot m'$ modulo p.

(2) Write down the addition and the multiplication tables for $\mathbb{Z}(5)$ and $\mathbb{Z}(7)$.

(3) Consider the congruence classes $[0], [1], [2], [3], [4]$ modulo 5. Solve the following equations: (a) $[2]+x = [0]$, (b) $[3]+x = [1]$, (c) $[3] +x = [4]$, (d) $[2] \cdot x = [0]$, (e) $[3] \cdot x = [1]$, and (f) $[3] \cdot x = [4]$. Solve the same equations over the congruence classes modulo 8.

(4) Consider Fermat's Little Theorem. In the proof, from the following equality

$$1 \cdot 2 \cdot \ldots \cdot (p-1) \equiv a \cdot 2a \cdot \ldots (p-1) \cdot a \equiv a^{p-1} \cdot (1 \cdot 2 \cdot \ldots \cdot (p-1)) \ (mod \ p)$$

we derived:

$$a^{p-1} \equiv 1 \ (mod \ p).$$

Prove the correctness of this step.

(5) Consider the proof of **Euler's Theorem**.

(a) Prove that no two distinct numbers in the sequence $(\star)$ are congruent modulo p.

(b) In the proof of the theorem, from the equality

$$(a_1 \cdot a_2 \cdot \ldots \cdot a_n) \equiv (a_1 \cdot a) \cdot (a_2 \cdot a) \cdot \ldots \cdot (a_n \cdot a) \equiv a^n \cdot (a_1 \cdot a_2 \cdot \ldots \cdot a_n) \ (mod \ m).$$

we derived that

$$a^{\phi(n)} = 1 \ (mod \ m).$$

Prove the correctness of this step.

(6) Run the RSA encryption on the following inputs:

 (a) $p = 7$, $q = 11$ and $M = 44$.
 (b) $p = 23$, $q = 41$, and $M = 35$.
 (c) $p = 19$, $q = 31$ and $M = 111$.

(7) Consider the RSA encryption method. Suppose that someone wants to decode the message C sent to you by your friend without knowing the prime numbers p and q (that constitute your private key). Discuss whether or not decoding is a complicated task.

Programming exercises

(1) Write a program that given positive integers c, d, n as inputs, computes c^d modulo n.
(2) Write a program that given positive integers e and n as inputs, outputs d such that $e \cdot d \equiv 1 \ (mod \ n)$.
(3) Implement the RSA encryption algorithm. It should support the encryption of text, and not just integers. It should have encode and decode functionality.

Lecture 6

Basics of directed graphs

A picture is worth a thousand words.
Fred R. Barnard.

6.1 Directed graphs and their representations

Graphs are used in all areas of computer science ranging from artificial intelligence and theoretical computer science to operating systems, networking and databases. Many computational problems in mathematics, biology, chemistry and other fields of science can be modeled as problems about graphs. Therefore, graphs are of fundamental importance in computer science, mathematics and their applications. In the next few lectures, we study graphs and some of the basic algorithms for graphs. We start with the definition of a directed graph.

Definition 6.1. A **directed graph** (or **digraph** for short) consists of points and directed lines connecting the points. Points are usually called **vertices** and the directed lines are called **edges**.

From this definition we see that in order to describe a directed graph, one needs to specify (1) its vertices and (2) its edges. Usually the collection of all vertices of the digraph is denoted by the upper case letter V, and the collection of all edges by the letter E. Digraphs can often be represented pictorially. Consider, as an example, the digraph in Figure 6.1.

The digraph in Figure 6.1 has 10 vertices represented as

$$V = \{0,\ 1,\ 2,\ 3,\ 4,\ 5,\ 6,\ 7,\ 8,\ 9\}.$$

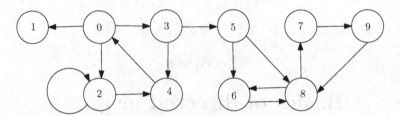

Fig. 6.1: An example of a directed graph

The edges in this digraph can be represented as ordered pairs of vertices. For example, the edge from 2 to 4 is represented as the pair $(2,4)$. There is no edge from 4 to 2, and hence the pair $(4,2)$ is not included in our description. We write the collection E of edges as

$$E = \{(0,1), (0,2), (0,3), (2,2), (2,4), (3,4), (3,5), (4,0), (5,6), (5,8),$$
$$(6,8), (7,9), (8,6), (8,7), (9,8)\}.$$

Thus, we see that we can translate a picture of a digraph into a written description, which enumerates all its vertices V and all its edges E. Given a set V of vertices and its edges E, we can also draw the corresponding digraph. For example, consider the digraph whose set V of vertices is

$$V = \{a, b, c, d, e, f\},$$

and whose set E of edges is

$$E = \{(a,b), (a,c), (a,d), (b,c), (b,d), (c,a), (c,f), (d,c), (e,a), (e,b), (f,d)\}.$$

A pictorial presentation of this digraph is in Figure 6.2.

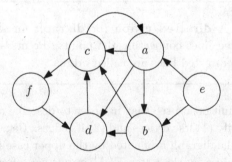

Fig. 6.2: An example of a directed graph

We denote digraphs with capital letters G, H, etc. From the definition of the digraph and the examples above, we conclude that we can write a directed graph G as $G = (V, E)$, where V is the collection of all vertices and E is the collection of all edges in the digraph. We note that we allow loops in digraphs. Namely, a loop (v, v) is allowed in digraphs. Pictorially, this is drawn as a looping arrow from vertex v to itself. For instance, see vertex 2 in Figure 6.1. If (v_1, v_2) is an edge of the digraph G then we say that the vertices v_1 and v_2 are **adjacent** to each other. In this case, the edge (v_1, v_2) is an **outgoing edge** from v_1 and an **incoming edge** to v_2. In some textbooks edges are also called *arcs* but for consistency we use our terminology.

Digraphs are frequently used in computer programs to represent real-world structures. For example, a flight planner program may represent all airports and flights as a digraph where the airports are represented as vertices and the direct flights are represented as edges between airports. For example, there is an edge from the Los Angeles vertex to the Seattle vertex because there is a direct flight. However, there is no edge between the Seattle vertex and the Sydney vertex since there is no direct flight between them.

There are some typical operations that may be performed on digraphs. These include inserting and deleting vertices, as well as inserting and deleting edges. For example, if there is a new direct flight from one city to another, say from Seattle to Sydney, a new directed edge must be inserted into the flight planner program. It is important to support these operations efficiently. Thus, we must carefully consider how we represent the digraphs, depending on what kind of operations we would like to support.

There are two popular ways to represent digraphs in computer programs (as data structures). The first is called the *adjacency list representation*. The second is the *adjacency matrix representation*. We explain these below.

For a given digraph $G = (V, E)$, the adjacency list representation is the following. For each vertex u in V, we list all v in V such that (u, v) is an edge of G. For example, the adjacency list representation for the digraph G in Figure 6.1 is

0 : 1, 2, 3.
1 :
2 : 2, 4.
3 : 4, 5.
4 : 0.
5 : 6, 8.

6 : 8.
7 : 9.
8 : 6, 7.
9 : 8.

Similarly, the adjacency list representation for the digraph G in Figure 6.2 is the following:

a : b, c, d.
b : c, d.
c : a, f.
d : c.
e : a, b.
f : d.

Thus, the adjacency list representation consists of lists. The number of lists is equal to the number of vertices of G. The list associated with vertex u consists of all the vertices v where (u, v) is an edge of G. If u has no outgoing edges then the list associated with u is the empty list.

In the adjacency matrix representation, the edge relation E is represented as a matrix. Suppose the vertices of G are 0, 1, ..., n. The digraph G is represented as a two-dimensional array such that the $(u, v)^{th}$ cell of the array has 1 if (u, v) is an edge in G, and has 0 otherwise. For example, the adjacency matrix representation for digraph G in Figure 6.1 is this:

$$\begin{pmatrix}
0 & 1 & 1 & 1 & 0 & 0 & 0 & 0 & 0 & 0 \\
0 & 0 & 0 & 0 & 0 & 0 & 0 & 0 & 0 & 0 \\
0 & 0 & 1 & 0 & 1 & 0 & 0 & 0 & 0 & 0 \\
0 & 0 & 0 & 0 & 1 & 1 & 0 & 0 & 0 & 0 \\
1 & 0 & 0 & 0 & 0 & 0 & 0 & 0 & 0 & 0 \\
0 & 0 & 0 & 0 & 0 & 0 & 1 & 0 & 1 & 0 \\
0 & 0 & 0 & 0 & 0 & 0 & 0 & 0 & 1 & 0 \\
0 & 0 & 0 & 0 & 0 & 0 & 0 & 0 & 0 & 1 \\
0 & 0 & 0 & 0 & 0 & 0 & 1 & 1 & 0 & 0 \\
0 & 0 & 0 & 0 & 0 & 0 & 0 & 0 & 1 & 0
\end{pmatrix}$$

Similarly, the adjacency matrix representation of the digraph G in Figure 6.2 is this:

$$
\begin{pmatrix}
0 & 1 & 1 & 1 & 0 & 0 \\
0 & 0 & 1 & 1 & 0 & 0 \\
1 & 0 & 0 & 0 & 0 & 1 \\
0 & 0 & 1 & 0 & 0 & 0 \\
1 & 1 & 0 & 0 & 0 & 0 \\
0 & 0 & 0 & 1 & 0 & 0
\end{pmatrix}
$$

Let us compare these two representations of digraphs. If G has n vertices then the digraph has *at most* n^2 edges. The size of the adjacency matrix representation is also n^2. Hence, when the digraph is *dense*, that is, the number of edges of G is close to n^2, both representations take a similar amount of space. When the digraph is *sparse*, that is, when the number of edges is close to n, the adjacency list representation saves space. Additionally, in the adjacency list representation, accessing *all neighbors* of a vertex u, that is all v such that (u, v) is in E, is fast. In the matrix representation accessing all neighbors of u takes time proportional to n. Indeed, in this case one needs to run through the uth row in the matrix and output all v such that the cell (u, v) in the matrix has 1.

However, given vertices u and v, checking whether (u, v) is an edge, in the adjacency list representation, takes time proportional to the number of edges outgoing from u. In contrast, in the adjacency matrix representation this takes a constant amount of time. Thus, in the flight planner example, if we want to quickly check if there is a direct flight possible from Los Angeles airport to Seattle airport, we can do this in constant time with the matrix representation. However, if we used the adjacency list representation, we would need to iterate through every single airport that has a direct flight out of Los Angeles and check if Seattle appears in this list. Thus, in this case, it may be better to use the matrix representation. These types of considerations must be taken into account when implementing graph algorithms. Different applications call for different representations. Thus, to maximize performance, one must carefully consider the application-specific requirements to decide which representation is most suitable.

6.2 Examples of digraphs

We present several examples of digraphs. Most of our examples have pictorial presentations. In these pictures, we represent vertices as circles and we do not name the vertices.

Example 6.1 (Directed cycles). *A pictorial presentation of the first five directed cycles is in Figure 6.3. The first directed cycle C_1 is a trivial cycle.*

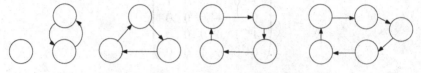

Fig. 6.3: Examples of directed cycles

The second C_2 is a directed cycle of length 2. The third C_3 is of length 3, etc.

Formally a directed cycle C_n of length n is defined as follows. The vertices of C_n are 0, 1, ..., $n - 1$. The edges are $(0, 1)$, $(1, 2)$, ..., $(n - 2, n - 1)$, $(n - 1, 0)$.

Example 6.2 (Directed wheels). *A pictorial presentation of the first three directed wheels is in Figure 6.4. The first directed wheel has four*

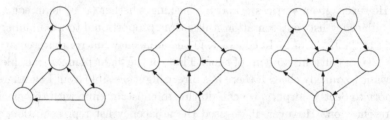

Fig. 6.4: Examples of directed wheels

vertices. The second has five vertices, etc.

Formally a directed wheel W_n, where $n > 2$, has $n+1$ vertices and is defined as follows. The vertices of W_n are 0, 1, ..., $n - 1$, and n. The edges are $(1, 2)$, $(2, 3)$, ..., $(n - 1, n)$, $(n, 1)$, and $(1, 0)$, $(2, 0)$, ..., $(n, 0)$.

Example 6.3 (Directed grids). *A pictorial presentation of three two dimensional directed grids is in Figure 6.5. The first directed grid has four vertices. The second has eight vertices. The third has twelve vertices, etc.*

Each directed grid is defined by two parameters n and m, where n indicates the length and m indicates the height of the grid. Therefore, we write directed grids as $Grid_{n,m}$. For example, the length of the third grid in

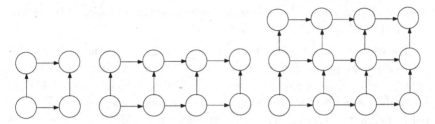

Fig. 6.5: Examples of directed grids

Figure 6.5 is 3 and the height is 2, hence the notation for that grid is $Grid_{3,2}$. Formally, a directed grid $Grid_{n,m}$ is described as follows. Its vertices are ordered pairs of the form (i, j), where $i = 0, \ldots, n$ and $j = 0, \ldots, m$. Hence the directed grid $Gird_{n,m}$ has $(n + 1) \cdot (m + 1)$ vertices. Its edges are of the form $((i, j), (i, j + 1))$ if $j < m$ and $i \leq n$, as well as edges of the form $((i, j), (i + 1, j))$ if $i < n$ and $j \leq m$.

Example 6.4 (Directed cubes). *A pictorial presentation of the first three directed cubes is in Figure 6.6. The first directed cube $Cube_1$ has*

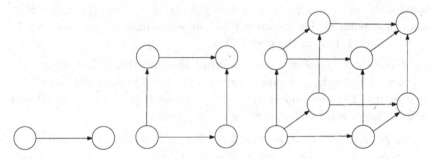

Fig. 6.6: Examples of directed cubes

two vertices. The second $Cube_2$ has four vertices. The third directed $Cube_3$ has eight vertices.

To formally define the directed cube $Cube_n$, we first give a more formal definition for the directed cubes $Cube_1$, $Cube_2$ and $Cube_3$.

The first directed cube $Cube_1$ can be viewed as an one-dimensional cube. Its vertices are 0 and 1, and it has only one edge which is $(0, 1)$. The second directed cube $Cube_2$ can be viewed as a two-dimensional cube. Its vertices are $(0, 0)$, $(1, 0)$, $(0, 1)$, $(1, 1)$. These vertices can be viewed as points in the plane. The edges between these vertices are $((0,0),(0,1))$, $((0,0),(1,0))$, $((1,0),(1,1))$, $((0,1),(1,1))$.

The directed cube $Cube_3$ can be viewed as a three-dimensional cube. Its vertices are the points P with coordinates (x,y,z) such that x, y and z are either 0 or 1. Thus, there are 8 vertices. To describe the edges of $Cube_3$, we need to recall how we add points in the 3-dimensional space. Given two points (x_1,y_1,z_1) and (x_2,y_2,z_2) in the space, the addition of these points produces the point whose coordinates are $(x_1+x_2,y_1+y_2,z_1+z_2)$. We write this:

$$(x_1,y_1,z_1)+(x_2,y_2,z_2)=(x_1+x_2,y_1+y_2,z_1+z_2).$$

For instance, the addition of $(1,2,-3)$ and $(2,-4,7)$ produces the point $(3,-2,4)$.

Now consider the following points that we call the *base points* in the 3-dimensional space:

$$B_1=(1,0,0), \quad B_2=(0,1,0), \quad \text{and } B_3=(0,0,1).$$

Using these base points we can describe the edges of the directed cube $Cube_3$ as follows. There is an edge from a vertex (x_1,y_1,z_1) of the cube to another vertex (x_2,y_2,z_2) if and only if $(x_2,y_2,z_2)=(x_1,y_1,z_1)+B$ for some base point B. For instance, there are edges from the vertices $(0,1,1)$, $(1,0,1)$ and $(1,1,0)$ to the vertex $(1,1,1)$.

Here is now a general description of the directed cube $Cube_n$. The vertices of the cube are points $(x_1,\ldots,x_n)$ in n-dimensional space such that each of the coordinates x_1, ..., x_n is either 0 or 1. There are 2^n such points. Now consider the following n base points:

$$B_1=(1,0,\ldots,0), \quad B_2=(0,1,0,\ldots,0), \quad \ldots, \quad \text{and } B_n=(0,\ldots,0,1).$$

Using these base points, we can describe the edges of $Cube_n$ as follows. There is an edge from a vertex $(x_1,x_2,\ldots,x_n)$ of the directed cube to another vertex $(y_1,y_2,\ldots,y_n)$ if and only if $(y_1,y_2,\ldots,y_n)=(x_1,x_2,\ldots,x_n)+B$ for some base point B. For instance, there are edges to the vertex $(1,1,\ldots,1)$ from all of the following n vertices:

$$(0,1,1,1,\ldots,1), \quad (1,0,1,1,\ldots,1), \quad (1,1,0,1,\ldots,1), \quad \ldots, \quad (1,\ldots,1,0).$$

6.3   Paths and strongly connected components in digraphs

Suppose we have a digraph $G=(V,E)$. We put a place-marker on a vertex v of the digraph and start moving the place-marker along the edges while respecting the direction of the edges. The place-marker travels and produces a path on the digraph. Along this path we can visit a vertex that has already been visited, we might also use an edge that has already been used. This is an informal explanation of the notion of a path in G. Here is a formal definition.

Definition 6.2. A **path** in a digraph $G = (V, E)$ is a sequence of vertices connected by edges so that the directions are respected. More formally, a path is a sequence

$$v_0, v_1, \ldots, v_i, v_{i+1}, \ldots, v_n$$

of vertices in G such that (v_0, v_1), ..., (v_i, v_{i+1}), ..., (v_{n-1}, v_n) are all edges in the digraph G. The **length** of this path is n. The vertex v_0 is the **start** of the path and v_n is the **end** of the path. We say that this path is from vertex v_0 to vertex v_n.

In the digraph G in Figure 6.1, examples of paths are: (*a*) 3, 5, 6, 8, 7, 9, and (*b*) 8, 7, 9, 8, 6. The lengths of these paths are 5 and 4, respectively. The sequence 4, 0, 3, 0, 2 is not a path since $(3, 0)$ does not constitute an edge. A path with no repeated vertices is called a **simple path**. Example (*a*) is a simple path, and example (*b*) is not a simple path. Note that in any given digraph, every vertex v forms a path on its own. The length of this path v is 0. Its start and end vertices are both v. A path is a **cycle** if it starts and ends with the same vertex. A cycle is **simple** if the only repeated vertex in the cycle is its start vertex, and it occurs in the cycle only twice. In the digraph G of Figure 6.1 the following paths are cycles: 3, 4, 0, 3 and 0, 2, 4, 0, 3, 4, 0. The first cycle is simple but the second cycle is not. The directed cycles C_n are simple cycles. In some textbooks, paths are called *walks* but we do not use this terminology.

We now would like to discuss connectivity in digraphs. Connectivity in digraphs is defined in terms of paths. Connectivity is one of the foundational definitions for digraphs.

Definition 6.3. We say that two vertices u and v of a digraph G are **strongly connected** if there exists a path from u to v and there exists a path from v to u.

The reader can try out this definition on the digraphs in Figures 6.1 and 6.2. For example, in Figure 6.1, the vertices 6, 7, 8, and 9 are all strongly connected to one another. However, the vertices 4 and 1 are *not* strongly connected. Although there is a path from vertex 4 to vertex 1, there is no path from 1 to 4. The next theorem states several important properties of this notion.

Theorem 6.1. *Let $G = (V, E)$ be a digraph. Each of the following is true:*

(1) Every vertex v is strongly connected to itself.

(2) If a vertex u is strongly connected to a vertex v then v is strongly connected to u.

(3) If a vertex u is strongly connected to a vertex v and v is strongly connected to w, then u is strongly connected to w.

Proof.

The proof of this theorem follows simply from the definition. The first part is true as every node v is a path (of length 0). Therefore, v is strongly connected to itself. From Definition 6.3, it is clear that the second part is also true. We prove the last part. Since u and v, and v and w are strongly connected there exist paths $v_0, \ldots, v_n$, and $w_0, \ldots, w_k$ such that the first path is from u to v and the second path is from v to w. We want to show that there exists a path from u to w. We put these two paths together and form the following sequence: $v_0, \ldots, v_n, w_1, \ldots, w_k$. Since $u = v_0$, $v = v_n = w_0$, and $w = w_k$, the sequence provided is a path starting with u and ending with w. A path from w to u is built in a similar way. Thus, u and w are strongly connected. $\qquad\square$

We write $u \sim v$ to indicate that vertices u and v are strongly connected. Then the properties stated in the theorem above can be restated as follows. For all vertices u, v, w of any given digraph G we have the following properties: (1) $v \sim v$, (2) if $u \sim v$ then $v \sim u$, and finally, (3) if $u \sim v$ and $v \sim w$ then $u \sim w$. These properties are analogous to the three properties of the modulo p congruence relation (that we denoted by $\equiv (mod\ p)$) studied in the previous lecture (See Theorem 4.2). In other words, strong connectedness between vertices in digraphs is reflexive, symmetric, and transitive.

Let v be a vertex of a digraph $G = (V, E)$. Consider *all* vertices u in G that are *strongly connected to* v. Let us list these vertices as:

$$u_1, \ u_2, \ \ldots, \ u_k.$$

From part (1) of the previous theorem, v must be among these vertices. Additionally, from the same theorem, if we take any two of these listed vertices, then they are strongly connected to each other. Moreover, if a vertex u is *not* among this list, then *no* vertex in this list is strongly connected to u. Indeed, if u were strongly connected to some u_i then, again by the theorem above, u would also be strongly connected to v. Therefore, u would have to appear among $u_1, \ u_2, \ \ldots, \ u_k$, which is a contradiction. Thus, keeping this observation in mind, we give the following definition:

Definition 6.4. Let G be a digraph and v be its vertex. The **strongly connected component of** v, which we denote by $C(v)$, is the collection of all vertices strongly connected to v. If $u_1, \ u_2, \ \ldots, \ u_k$ are all the vertices that are strongly connected to v, then we write this fact as

$$C(v) = \{u_1, u_2, \ldots, u_k\}.$$

We also say that $C(v)$ is the strongly connected component produced by the vertex v.

We say that a digraph is **strongly connected** if it consists of exactly one strongly connected component. This is equivalent to saying that every pair of vertices of the digraph is strongly connected.

It is easy to verify, using the theorem above, that if two vertices, say u and v, are strongly connected, then they produce the same strongly connected component, i.e. $C(u) = C(v)$. As an example, consider the digraph G in Figure 6.1. The list

$$C(0) = \{0, 2, 3, 4\}, \ C(1) = \{1\}, \ C(5) = \{5\}, \ C(6) = \{6, 7, 8, 9\}$$

exhausts *all* strongly connected components of the digraph G. Note that we have omitted the strongly connected component of the vertex 2 (for instance) from the list, since the vertices 2 and 0 produce the same strongly connected component.

6.4   Exercises

(1) Consider the following digraph $(\{0, 1, 2, 3, 4, 5, 6, 7, 8, 9\}, E)$, where $E = \{(n, m) \mid n \equiv m \ (mod\ 3)\}$. Draw the digraph and write down the adjacency matrix and adjacency list representations for this digraph.

(2) Consider the digraph $G = (V, E)$, where:

  (a) The set V consists of points (x, y) in the plane such that the coordinates x and y are natural numbers ≤ 4.

  (b) The set of E edges consists of pairs of points $((x_1, y_1), (x_2, y_2))$ such that either $(x_1 + 1 = x_2$ and $y_1 = y_2)$ or $(y_1 + 1 = y_2$ and $x_1 = x_2)$ or $(x_1 + 1 = x_2$ and $y_1 + 1 = y_2)$ or $(x_1 = y_1 = 4$ and $x_2 = y_2 = 0)$.

  Do the following: (a) Draw the digraph. (b) How many entries does the adjacency matrix for the digraph have? (c) Is the digraph strongly connected?

(3) Let $G = (V, E)$ be a digraph. Prove the following properties of the strongly connected components of G:

  (a) Every vertex belongs to some strongly connected component of G.

  (b) Consider two strongly connected components, say $C(u)$ and $C(v)$. Then either $C(u)$ and $C(v)$ do not share any vertex or $C(u)$ and $C(v)$ are the same strongly connected components.

(4) Let $G = (V, E)$ be a digraph. For each $v \in V$, the **in-degree of** v is the number of incoming edges to v, and the **out-degree of** v is the number of outgoing edges from v. Prove that the sum of all in-degrees is equal to the sum of all out-degrees.

(5) Consider the following digraph $G = (V, E)$. The set V of vertices are all natural numbers: $0, 1, 2, \ldots$. The edges are $(3n, 3n+1)$, $(3n+1, 3n+2)$, and $(3n + 2, 3n)$, where n ranges over all natural numbers. Do the following:

  (a) Write down the strongly connected components of vertices 10, 17, and 30.

  (b) Describe the strongly connected components.

  (c) How many strongly connected components does G have?

(6) Write down the adjacency list and matrix representations of directed wheels and directed cycles (see Examples 6.1 and 6.2).

(7) Consider a vertex v in the directed cube $Cube_n$ (See Example 6.4). How would you compute the in-degree and the out-degree of the vertex v?

(8) What is the length of the longest path in the directed cube $Cube_n$?

Programming exercises

(1) Implement the following two methods:

    (a) the first takes the adjacency matrix representation of a digraph G and outputs the adjacency list for G, and

    (b) the second takes the adjacency list representation of a digraph G and outputs the adjacency matrix for G.

(2) Write a method that takes a digraph G and a vertex v in the digraph, and outputs the strongly connected component $C(v)$.

(3) Write a program that, given a digraph G as input, outputs all the vertices of even out-degree.

(4) Write a program that, given a digraph G as input, outputs the sum of the in-degrees of all the vertices.

Lecture 7

The path problem and undirected graphs

Two roads diverged in a wood, and –
I took the one less traveled by,
And that has made all the difference.
Robert Frost.

7.1 The path problem

The **path problem** is stated as follows:

Given a digraph G and its vertices s and t, *determine* whether there exists a path from s to t.

Our goal in this lecture is to find an efficient algorithm that solves the path problem. In other words, we want an algorithm that, given a digraph G and vertices s and t, outputs *"yes"* if there is a path from s to t, and *"no"* otherwise. The vertex s is called the **source** vertex, and the vertex t is called the **target** vertex.

The path problem and many of its variations are common in most areas of computer science such as artificial intelligence, verification and model checking, databases, graphics, robotics, and so on. For example, in model checking the goal is to determine whether a system satisfies a given set of requirements. The system might be a hardware system or your favorite application program. Often, the system is represented using digraphs, where the vertices represent different states of the system and edges represent transitions between the states. One way to think about the states is that some states are "good" states (i.e. the system is running normally) and some states are "bad" states (i.e. error states). A common requirement in many systems is the following. Though the system is allowed to reach "bad" states, there should always be a path from any "bad" state back into a "good" state. Clearly, this can be restated as the path problem.

We begin with an algorithm, called a *brute-force* algorithm, that solves the path problem. The name *brute-force* comes from the fact that the algorithm examines *all* potential paths to determine if there exists a path from s to t. We say that a *potential path* of length k is any sequence $v_0, \ldots, v_k$ of vertices from V. Of course, not every potential path is a path. However, given a potential path, it is easy to determine whether it is indeed a path; one just needs to verify that (v_i, v_{i+1}) is an edge of the graph for all $i = 0, \ldots, k - 1$.

Let G be a digraph with n vertices. The number of potential paths of length 0 is n; the number of potential paths of length 1 is n^2; the number of potential paths of length 2 is n^3, etc. Every potential path of length $k - 1$ can be extended to a potential path of length k in exactly n different ways. Therefore, one can observe that the number of potential paths of length k is n^{k+1}. (This type of reasoning can be made precise by using the induction principle which we will learn in forthcoming lectures.)

The brute-force algorithm examines all potential paths of length up to n. For each potential path, the algorithm verifies (1) whether the potential path starts with s and ends with t, and (2) whether the potential path constitutes a path. If the algorithm finds such a path, it stops by reporting that a path from s to t is found (i.e. it ouputs *"yes"*). Otherwise, the algorithm runs through all potential paths. Once the algorithm exhausts all the potential paths, it then stops by reporting that a path from s to t does not exist (i.e. it ouputs *"no"*). The brute-force algorithm is *correct*. However, the algorithm is very *inefficient*. In the worst case, it needs to examine all potential paths of length 0, 1, 2, $\ldots$, $n - 1$, which is clearly not feasible for large n.

We now provide a more efficient algorithm, the Algorithm 7.1 below, which we call the *PathExistence*(G, s, t) algorithm. Inputs to this algorithm are digraphs G and vertices s and t of G. The output is *"yes"* if a path from s to t exists, and *"no"* otherwise.

This algorithm can be efficiently implemented in many ways. We do not discuss issues related to its implementation. Rather, our goal is to prove that the algorithm is correct. However, the interested reader is welcome to implement the algorithm for Programming Exercise 1.

Here are some observations about the algorithm. Let n be the number of vertices in the digraph G. Lines 1 and 3 are executed exactly once. In line

---

Algorithm 7.1 *PathExistence*(G, s, t) algorithm

---

(1) Mark s as visited.
(2) Repeat until no more vertices can be marked:

    (a) Iterate through all edges. If an edge (u, v) is found, such that u is marked and v is not, then mark v as visited.

(3) If t is marked visited then report *yes* and stop. Otherwise, report *no* and stop.

$2a$, the algorithm runs through all the edges. Each time it runs through all the edges, apart from the last time, it marks at least one vertex as visited. Once a vertex is marked as visited, the vertex will never be marked again. Therefore, line $2a$ of the algorithm is executed at most n times. Once the algorithm executes line $2a$ for the last time, it goes to line 3 and stops.

To show that the algorithm is *correct*, we need to verify the following two properties:

(1) If t is marked visited, then there is a path from s to t.
(2) If there is a path from s to t, then t is marked visited.

We prove the first part. Suppose that line $2a$ of the algorithm is executed exactly m times. As noted above, $m \leq n$. We can also suppose that every time line $2a$ is executed, apart from the last one, it marks *exactly* one vertex as visited. So, let

$$s, v_1, v_2, \ldots, v_m$$

be all the vertices marked in order of executions of line $2a$. Note that s is marked in line 1 at the beginning. Clearly, there is an edge from s to v_1. Similarly, there is an edge from either s to v_2 or from v_1 to v_2. Therefore, there is a path from s to v_2. We can continue this reasoning vertex-by-vertex. Now suppose we have considered every vertex in the sequence s, $v_1, \ldots, v_i$, where $i \leq m - 1$. We can assume that we have proved that there is a path from s to each of the vertices from $v_1, \ldots, v_i$. Since v_{i+1} is marked, there must exist an edge from one of the vertices $v_1, \ldots, v_i$ to v_{i+1} according to line $2a$ of the algorithm. Say that this vertex is v_j. Thus, there exists a path from s to v_j and there is an edge from v_j to v_{i+1}. Therefore, there must exist a path from s to v_{i+1}. Hence, for every marked vertex v there is a path from s to v. Thus, if t is marked, then there is a path from s to t. We have proved the first part.

To prove the second part, suppose that there is a path $w_0, w_1, \ldots, w_k$ from s to t. Thus, $w_0 = s$ and $w_k = t$. The vertex w_0 is marked. Therefore, line 2a must mark w_1 at some later stage. Vertex w_1 is thus marked. By the same reasoning, line 2a must also mark w_2 at some stage. Continue this on, to consider the path $w_0, w_1, \ldots, w_i$, where $i \leq k - 1$. We can assume that we have proved that all the vertices $w_0, \ldots, w_{i-1}$ are marked. Since w_{i-1} is marked and (w_{i-1}, w_i) is an edge, the vertex w_i must also be marked as instructed by line 2a of the algorithm. Thus, $t = w_k$ will be marked. Therefore, we have proved the second part, and have finished proving the correctness of the $PathExistence(G, s, t)$ algorithm.

7.2 Undirected graphs

In digraphs, the edges have directions indicated by arrows. In undirected graphs, the edges do *not* have directions. Here is a more formal definition:

Definition 7.1. An **undirected graph** consists of points and lines between the points. Points are called **vertices** and lines are called **edges**. No edge in an undirected graph connects a vertex to itself. We also do not allow more than one edge between any two vertices. Often, the word **graph** will mean an undirected graph.

A natural example of a graph is a social network graph; each person is represented by a vertex and there is an edge between two vertices, say between Amy and Bob, if Amy and Bob are friends. Such graphs are maintained by social networking websites so that the website can show only relevant information to users (i.e. users are interested in seeing only information related to themselves and those they know), as well as to protect the privacy of users (i.e. a stranger should not be able to view a user's personal information).

Thus, every undirected graph G consists of the set V of vertices and the set E of undirected edges. Therefore, G can be written as the pair (V, E). As for directed graphs, we can represent graphs pictorially. Consider, for example, the graph presented in Figure 7.1.

The set V of vertices of this graph is:

$$V = \{0, 1, \ldots, 9\}.$$

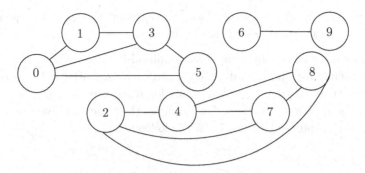

Fig. 7.1: An example of a graph

To denote edges of this graph we need to be a little careful. Consider for instance the edge between 0 and 5. We cannot denote this edge by $(0, 5)$ since $(0, 5)$ is used to indicate the directed edge from 0 to 5. For the same reason we can not denote this edge by the pair $(5, 0)$. Instead, we denote the edge between 0 and 5 by $\{0, 5\}$. Since this edge does not have a direction $\{5, 0\}$ represents the same edge. So, the set E of edges can be written as

$$E = \{\{0, 1\}, \{0, 3\}, \{1, 3\}, \{3, 5\}, \{0, 5\}, \{2, 4\}, \{2, 7\}, \{2, 8\}, \{4, 7\}, \{7, 8\},$$
$$\{4, 8\}, \{6, 9\}\}$$

Note E does not contain, for instance, $\{3, 1\}$ since it already contains $\{1, 3\}$ representing the same edge. Putting $\{3, 1\}$ into the edge set E would therefore be redundant.

Let G be a graph and e be an edge in G. Since e is an edge it connects two vertices v and u in G. As explained above, we represent the edge e as $\{u, v\}$. We again stress the difference between notations for edges in directed and undirected graphs. In digraphs the edges (u, v) and (v, u) are *different*, while in undirected graphs the edges $\{u, v\}$ and $\{v, u\}$ are *same*. Note also that every undirected graph can be viewed as a directed graph G by replacing every undirected edge $e = \{u, v\}$ in G with two directed edges (u, v) and (v, u).

As for directed graphs, there are some typical operations that may be performed on graphs. These might include inserting and deleting vertices, as well as inserting and deleting edges. For instance, if a social networking service has a new user, a new vertex must be inserted into the graph, or if Amy does not want to be Bob's friend, then the edge between Amy and Bob must be deleted. We would like to support these operations efficiently. For this, we must carefully consider how we represent the graphs, depending on types operations we would like to support.

As for directed graphs, undirected graphs also have two representations: the adjacency list and the adjacency matrix representations. However, we can save space in these representations for undirected graphs due to the absence of directions on edges. For example, the adjacency list representation of the graph G in Figure 7.1 is the following:

```
0 :  1, 3, 5.
1 :  3.
2 :  4, 7, 8.
3 :  5.
4 :  7, 8.
6 :  9.
7 :  8.
```

We did not list all the nodes adjacent to 8. This is because 8 already appears in the list of neighbors of vertices 2, 4, and 7. Also, the list 4 : 7, 8 does not contain 2 because the vertex 4 appears in the list 2 : 4, 7, 8. Similarly, the adjacency matrix representation of the graph G can be written as follows where all the cells (i, j) below the diagonal of the matrix have 0.

$$
\begin{pmatrix}
0 & 1 & 0 & 1 & 0 & 1 & 0 & 0 & 0 & 0 \\
0 & 0 & 0 & 1 & 0 & 0 & 0 & 0 & 0 & 0 \\
0 & 0 & 0 & 0 & 1 & 0 & 0 & 1 & 1 & 0 \\
0 & 0 & 0 & 0 & 0 & 1 & 0 & 0 & 0 & 0 \\
0 & 0 & 0 & 0 & 0 & 0 & 0 & 1 & 1 & 0 \\
0 & 0 & 0 & 0 & 0 & 0 & 0 & 0 & 0 & 0 \\
0 & 0 & 0 & 0 & 0 & 0 & 0 & 0 & 0 & 1 \\
0 & 0 & 0 & 0 & 0 & 0 & 0 & 0 & 1 & 0 \\
0 & 0 & 0 & 0 & 0 & 0 & 0 & 0 & 0 & 0 \\
0 & 0 & 0 & 0 & 0 & 0 & 0 & 0 & 0 & 0
\end{pmatrix}
$$

A technical but important concept for graphs is the notion of degree. Let v be a vertex in a graph. Then the **degree** of v, written $deg(v)$, is the number of edges connected to v. For example, for graph G in Figure 7.1, $deg(4) = deg(7) = deg(0) = 3$ and $deg(5) = 2$.

7.3   Examples of graphs

In this section, we present several examples of graphs. For these examples, vertices are represented as circles and they are not named. The reader can test many notions and concepts for graphs in these examples. For instance, one can think about computing degrees of vertices for these examples of graphs, or writing down their adjacency matrix representations. Directed graph versions of some of these examples are presented in Section 6.2.

Example 7.1 (Complete graphs). *We depict four complete graphs in Figure 7.2. The first graph is a trivial one, the second is the cycle of length*

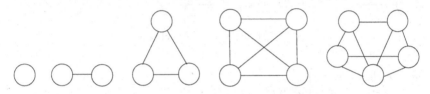

Fig. 7.2: Examples of complete graphs

2, the third is the cycle of length 3, the fourth is the graph with four vertices that are all connected with edges, etc.

Formally a complete graph with n vertices is denoted by K_n. It can be defined as follows. The vertices of K_n are $0, 1, \ldots, n-1$. The edges are of the form $\{i, j\}$ where $i = 0, \ldots, n-1$, $j = 0, \ldots, n-1$, and $i \neq j$.

Example 7.2 (Cycles). *A pictorial presentation of four cycles is in Figure 7.3. The first cycle is of length 2, the second cycle is of length 3, etc.*

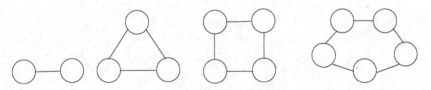

Fig. 7.3: Examples of cycles

Formally a cycle C_n of length n is defined as follows. The vertices of C_n are $0, 1, \ldots, n-1$. The edges are $\{0, 1\}, \{1, 2\}, \ldots, \{n-2, n-1\}, \{n-1, 0\}$. Note that we are using the same notation for directed and undirected cycles C_n.

It will always be clear whether we are referring to directed or undirected cycles, because we will explicitly refer to them as directed cycles or, simply, as cycles.

Example 7.3 (Wheels). *A pictorial presentation of the first three wheels is in Figure 7.4. The first wheel has four vertices. The second has five, and*

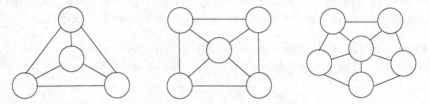

Fig. 7.4: Examples of wheels

the third has 6 vertices.

Formally, an undirected wheel W_n has $n + 1$ vertices and is defined as follows. The vertices of W_n are $0, 1, \ldots, n-1$, and n. The edges are $\{1, 2\}$, $\{2, 3\}, \ldots, \{n-1, n\}, \{n, 1\}$, and $\{1, 0\}, \{2, 0\}, \ldots, \{n, 0\}$.

Example 7.4 (Grids). *Grids are obtained from directed grids $G_{n,m}$ explained in Example 6.3 by removing the directions from edges. A pictorial presentation of grid $G_{7,2}$ is in Figure 7.5.*

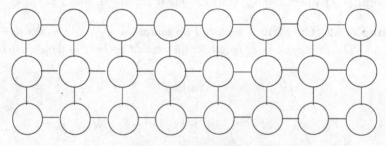

Fig. 7.5: Grid $G_{7,2}$

As for directed grids, each grid has two parameters n and m. The first parameter indicates the length of the grid, and the second parameter indicates the height of the grid. Therefore, we write directed grids as $Grid_{n,m}$. Formally, vertices of $G_{n,m}$ are of the form (i, j), where $i = 0, \ldots, n$ and

$j = 0, \ldots, m$. It has edges of the form $\{(i,j),(i,j+1)\}$ if $j < m$ and $i \leq n$, as well as edges of the form $\{(i,j),(i+1,j)\}$ if $i < n$ and $j \leq m$.

Example 7.5 (Cubes). *Cubes are obtained from directed cubes by removing the directions on edges. See Example 6.4. Again, as we did for cycles, wheels and grids, we use the same notation for cubes $Cube_n$ as for directed cubes.*

Here is now a more formal description of the cube $Cube_n$. The vertices of the cube are points $(x_1, \ldots, x_n)$ in the n-dimensional space such that each of the coordinates $x_1, \ldots, x_n$ is either 0 or 1. There are 2^n such vertices. Now consider the following n points in the space

$$B_1 = (1,0,\ldots,0), \ B_2 = (0,1,0,\ldots,0)), \ \ldots, \text{ and } B_n = (0,\ldots,0,1).$$

We called them, in Example 6.4, the *base points*. Using these base points we can describe the edges of $Cube_n$ as follows. There is an edge between a vertex $(x_1, x_2, \ldots, x_n)$ and a vertex $(y_1, y_2, \ldots, y_n)$ if and only if $(y_1, y_2, \ldots, y_n) = (x_1, x_2, \ldots, x_n) + B$ for some base point B. For instance, there are edges between the vertex $(1, 1, \ldots, 1)$ and all of the following n vertices:

$$(0,1,1,1,\ldots,1), \ (1,0,1,1,\ldots,1), \ (1,1,0,1,\ldots,1), \ \ldots, \ (1,\ldots,1,0).$$

The reader should notice the difference in description of edges of cubes as directed graphs (See Example 6.4) and cubes as undirected graphs.

7.4 Path and components in graphs

Many concepts for directed graphs make sense for undirected graphs. For example, a **path** is a sequence $v_0, v_1, \ldots, v_n$ of vertices such that $\{v_0, v_1\}$, $\{v_1, v_2\}$, $\ldots$, $\{v_{n-1}, v_n\}$ are all edges. If all the vertices in the path v_0, $v_1, \ldots, v_n$ are pairwise distinct then we call the path a **simple path**. Clearly, if $v_0, v_1, \ldots, v_n$ is a path then so is $v_n, v_{n-1}, \ldots, v_1, v_0$. This statement is not always true for digraphs. One can also define the notions of connectivity and components for graphs similar to strong connectivity and strongly connected components for directed graphs.

Definition 7.2. Two vertices u and v in graph G are **connected** if there is a path from u to v. The graph G is **connected** if all pairs of vertices u, v of the graph are connected.

For instance, for graph G in Figure 7.1 all the vertices 2, 4, 7 and 8 are connected. Now Theorem 6.1 from the previous lecture can be restated for undirected graphs.

Theorem 7.1. *Let G be a graph. Then each of the following is true:*

(1) Every vertex v of the graph is connected to itself.

(2) If a vertex v is connected to a vertex u then u is connected to v.

(3) If a vertex v is connected to u and u is connected to w then v is connected to w.$\square$

We leave the proof as Exercise 4 to the reader. Based on this theorem, we define components for graphs as follows. The **component** of vertex v, written as $C(v)$, is the collection of *all* vertices u that are connected to v. For example, graph G in Figure 7.1 has three components

$$C(0) = \{0, 1, 3, 5\}, \ C(2) = \{2, 4, 7, 8\}, \ \text{and} \ C(6) = \{6, 9\}.$$

Note that a connected graph has exactly one component, which is the graph itself.

7.5   Exercises

(1) Consider the graphs and digraphs in Figure 7.6 below:

    (a) Write down their adjacency list and adjacency matrix representations.

    (b) Find all (strongly) connected components of each graph.

(2) Prove that the brute-force algorithm for solving the path problem is correct.

(3) We say that a graph is k-**regular**, where k is a positive integer, if every vertex v in the graph has exactly k neighbors. For every positive *even* integer $n > 2$ construct a 3-regular graph with n vertices. (*Hint:* First, construct 3-regular graphs with 4, 6, and 8 vertices)

(4) Prove Theorem 7.1.

(5) Let G be an undirected graph. Prove that the sum of the degrees of the vertices of G is always an even number. (*Hint:* Discuss the contribution of each edge to the sum).

(6) Prove that the number of vertices of odd degree in any graph G is even.

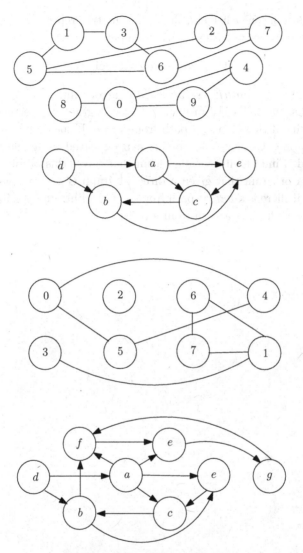

Fig. 7.6: Examples of graphs

(7) Consider the *PathExistence*(G, s, t) algorithm. Assume that checking if a vertex is marked takes one unit of time. If G has n vertices, then give an upper bound on time (in terms of time units) needed for the *PathExistence*(G, s, t) algorithm to stop. Discuss the efficiency of the algorithm.

(8) Explain adjacency list and adjacency matrix representations of complete graphs, cycles, wheels, grids, and cubes.

Programming exercises

(1) Implement the $PathExistence(G, s, t)$ algorithm.
(2) Consider the following problem. Given a graph G, vertices s and t, determine if there exists a path from s to t. If such a path exists then output one. Clearly this problem can be solved using the brute-force method. Find an efficient algorithm that solves this problem.
(3) Write a program that given a directed graph G and two vertices s, t, checks if there is a *unique* path from s to t. (Hint: one possible solution starts with finding a path from s to t.)

Lecture 8

Circuit problems in graphs

Those who cannot remember the past are condemned to repeat it.

George Santayana.

8.1 Euler circuits in graphs

Suppose we are visiting an ancient city (say Kyoto, Samarkand, or Rome). The city has many historical attractions. There are routes between the attraction sites. Our goal is to visit *all* the attractions by using *all* the routes, and *never repeating* the same route twice. This task can easily be modeled as a graph problem, if we view the attraction sites as vertices of the graph and the routes as edges between the vertices. We would like to formulate the above problem for an arbitrary graph. For this, we first need the definition below.

Definition 8.1. Suppose that we are given a graph $G = (V, E)$. We say that a path $v_0, v_1, \ldots, v_n$ is an **Euler circuit** if it satisfies each of the following properties:

(1) The first vertex v_0 is the same as the last vertex v_n, that is $v_0 = v_n$.
(2) Every edge of the graph occurs in the path.
(3) No edge in this path is repeated.
(4) Every vertex of the graph occurs in the path.

Observe that the definition does *not* prohibit the same vertex occurring in an Euler circuit more than once. Also note that any graph that has an Euler circuit must be connected. In addition, the length of an Euler circuit is equal to the number of edges of the graph. If we think of the vertices as the attraction sites of a city, and view the edges as routes between the attraction sites, then the existence of an Euler circuit tells us that it is possible to tour *all* the attractions by using *all* the routes, *never repeating the same route twice* and returning to the attraction from which the tour started. For example, consider the graphs in Figure 8.1. Which of these graphs have Euler circuits and which do not?

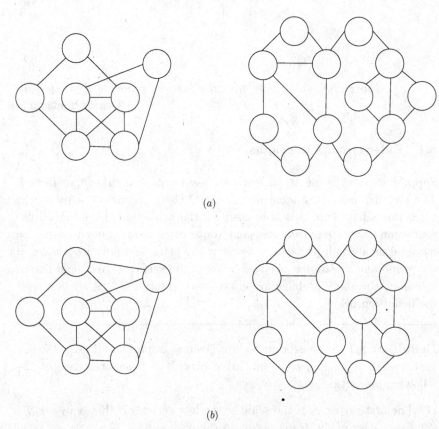

(a)

(b)

Fig. 8.1: Which of these graphs have Euler circuits?

The **Euler circuit problem** is stated as follows:

Given a graph $G = (V, E)$ determine whether G has an Euler circuit.

Just like the path problem from the previous lecture, there is a brute-force algorithm that solves the Euler circuit problem. The brute-force algorithm proceeds as follows. Say e_1, e_2, ..., e_m are *all* the edges of the graph. The algorithm considers all possible listings of these edges, so that no edge is repeated. Then the algorithm searches for an Euler circuit in this space of *all* listings. Once such a list is detected, the algorithm stops and reports that an Euler circuit exists. Otherwise, after searching though the whole space of listings, the algorithm reports that no Euler circuit exists. This algorithm is *inefficient*. Therefore our goal is to investigate graphs with an eye towards finding a clever algorithm for solving the Euler circuit problem.

Suppose that $G = (V, E)$ is a graph with an Euler circuit. We would like to derive some properties of G, using the fact that G has an Euler circuit. For this, recall that the degree of a vertex v in graph G, written $deg(v)$, is the number of edges connected to v.

By our assumption, the graph G has an Euler circuit. Let us take an Euler circuit in G: v_0, v_1, ..., v_n. We walk along this path going from v_0 to v_1, from v_1 to v_2, and so on, and erase the edge that has just been passed. Each time we go through a vertex w in this walk we erase two edges: one is the edge going into w, and the other is the edge outgoing from w. Therefore, if the vertex w occurs in the Euler path k times and $w \neq v_0$, then we erase exactly $2k$ edges connected to w. Therefore, the degree of w is $2k$, since all edges must appear in the Euler circuit v_1, ..., v_n. For the case when $w = v_0$ (and thus $w = v_n$) and w occurs in this path t times, the number of edges connected to v_0 is $2t - 2$. We state this observation as the following theorem.

Theorem 8.1. *Every vertex in any graph with an Euler circuit has even degree.* $\square$

From this theorem, we know that the graphs in 8.1(b) do not have Euler circuits because they have vertices of odd degrees. Without knowing this theorem, it would have been harder to detect that these do not have Euler circuits.

We would like to note that determining whether the degree of every vertex in a graph is even or odd can be done efficiently. One just goes through every vertex in the adjacency (or matrix) list representation and counts the number of edges of the vertex.

Now, we converse the theorem above, the result proved by Leonhard Euler in 1736.

Theorem 8.2. *A connected graph whose vertices all have even degree has an Euler circuit.*

Proof.

Let $G = (V, E)$ be a connected graph in which every vertex has an even degree. We want to prove that G has an Euler circuit. This proof consists of two parts. First, we introduce the concept of an almost Euler circuit. Then we prove that every connected graph satisfying the hypothesis of the theorem must have a almost Euler circuit. Second, we use the concept of a almost Euler circuit to build an Euler circuit in the graph, thus proving the theorem.

Definition 8.2. We say that a path $v_0, v_1, \ldots, v_k$ with $k > 1$ in a graph G is an **almost Euler circuit** if it satisfies the following properties:

(1) The first vertex v_0 is the same as the last vertex v_k, that is $v_0 = v_k$.
(2) All the edges in the path $v_0, v_1, \ldots, v_{k-1}$ are pairwise distinct.

Clearly, an Euler circuit is an almost Euler circuit but the converse is not always true, since almost Euler circuits are not required to use all the edges nor to visit all the vertices. We now prove one nice fact about almost Euler circuits. The fact is then used to prove the theorem. A fact used in a proof of a theorem is often referred to as a **lemma**. For the lemma below, we recall that G is a connected graph whose vertices all have even degree.

Lemma 8.1. *For every vertex v_0 in G, there exists an almost Euler circuit starting with the vertex v_0.*

Indeed, let us take the vertex v_0. Take an arbitrary edge e starting with v_0 and move along the edge to the vertex v_1 (determined by e). This produces a path v_0, v_1. Remove the edge e from the graph. Since all the vertices had an even degree, in the new graph the degrees of v_0 and v_1 are odd and the degrees of all other vertices are even. Since the degree of v_1 is odd, it is not zero. So, we can select an edge $\{v_1, v_2\}$ and produce the path v_0, v_1, v_2. Remove this edge $\{v_1, v_2\}$ from the graph. We have used exactly two edges connected to v_1. Therefore, the number of remaining edges connected to v_1 is still even. Possibly, v_1 may have no edges left because we may have removed all its edges. In the new graph, the degrees of v_0 and v_2 are odd, and the degrees of all other vertices are even. Since the degree of v_2 is now odd, we can select an edge $\{v_2, v_3\}$ and produce the path v_0, v_1, v_2, v_3. Remove this edge $\{v_2, v_3\}$ from the graph. Again, we have used exactly two edges connected to v_2. Therefore, the number of remaining edges connected to v_2 is still even. If $v_3 = v_0$, then we have an almost Euler circuit and the lemma is proved. Otherwise, in the new graph, the degrees of v_0 and v_3 are odd and the degrees of all other vertices are even. We now repeat the process for v_3. Continue this process of building a path and removing the edges that have been used. Suppose now that we have built the following path:

$$v_0, v_1, v_2, \ldots, v_{i-1}, v_i.$$

Remove the last used edge (i.e. $\{v_{i-1}, v_i\}$). Again, we have used exactly two edges connected to v_{i-1}. Therefore, the number of remaining edges connected to v_{i-1} is still even.

Note the following. First, we never use the same edge twice because every edge is removed after it is used. Second, in this path, when we process a vertex, say w, where w is among $v_1, v_2, \ldots, v_{i-1}$, we reduce the degree of w by 2. This is because we remove the edge leading into it and the one leaving from it. Therefore, the degree of w in the new graph remains even. If $v_i = v_0$, then we have an almost Euler circuit and the lemma is proved. Otherwise, in the new graph the degrees of v_0 and v_i are odd and the degrees of all other vertices are even. Therefore, we can continue the process for v_i. The process of building a path must stop and it can not stop at any vertex other than the starting vertex v_0. The lemma is proved.

Now, we prove the theorem using the lemma above. Take a vertex v. Using the lemma, build an almost Euler circuit C_1, starting with v. If C_1 is a Euler curcuit, then we are done. Otherwise, consider the graph G_1, obtained from G, by removing all the edges that occur in the path C_1. We know that in G_1 every vertex has an even degree. Now, choose a vertex w in C_1 that has a non-zero degree in G_1. Such a vertex *must exist*. Apply the lemma again to graph G_1, to build an almost Euler circuit C_2 from w. These two almost Euler circuits, C_1 and C_2, can be attached to each other to form a longer almost Euler circuit C. As an example, if C_1 were $v_0, v_1, v_2, v_3, v_4, v_5$ and C_2 were $w_0, w_1, w_2, w_3, w_4, w_5, w_6, w_7$ with $v_2 = w_4$ then we would form the following longer almost Euler circuit

$$v_0, v_1, v_2, w_5, w_6, w_7, w_1, w_2, w_3, w_4, v_3, v_4, v_5.$$

If C is an Euler circuit then the theorem is proved. Otherwise, we rename C as C_1 and repeat the same reasoning applied to this new almost Euler circuit C_1. The process must stop and produce an Euler circuit.   □

Below we present an algorithm that outputs an Euler circuit in connected graphs. For this algorithm, we always assume that the input graphs are connected with *all vertices of even degree*. First, we need an algorithm that builds almost Euler circuits. Lemma 8.1 takes care of its correctness. Here is the *AlmostEuler*(G, v) procedure, where G is a connected graph:

Algorithm 8.1 *AlmostEuler*(G, v) algorithm

---

(1) Choose an edge e starting at v.

(2) Set P to be the path e, and remove e from the edge list of G.

(3) *While* the last vertex of P is not equal to v *do*

    (a) Choose an edge e_1 starting from the last vertex of P.

    (b) Adjoin e_1 to P (and so P is now a longer path).

    (c) Remove e_1 from the edge list of G.

(4) Output P

We use the *AlmostEuler*(G, v) procedure in the *EulerCircuit*(G) algorithm, which builds an Euler circuit. The correctness of the algorithm is provided by the two theorems above. The *EulerCircuit*(G) algorithm, where G is a connected graph, is presented below:

Algorithm 8.2 *EulerCircuit*(G) algorithm

---

(1) Choose a vertex v.

(2) Let $C_1 = $ *AlmostEuler*(G, v).

(3) *While* C_1 is not a Euler curcuit *do*:

    (a) Remove all edges of C_1 from the edge list of G. Set G_1 to be the remaining graph.

    (b) Choose a w in C_1 that has a positive degree in G_1.

    (c) Call *AlmostEuler*(G_1, w). Set C_2 to be the path produced by *AlmostEuler*(G_1, w).

    (d) Attach C_2 to C_1 to form a longer path.

    (e) Rename the new longer path C_1.

(4) Output C_1.

Theorems 8.1 and 8.2 that we have proved allow us to solve the Euler circuit problem of determining whether a given *connected* graph G has an Euler circuit. For this, one just needs to count the parity of the degrees of each vertex in the graph. This can be easily implemented by running through the adjacency list (or matrix) representation of the input graph. The algorithm needs to simply check if $deg(v)$ is even for every vertex v of G.

The assumption on connectivity in the theorems above is important. For example, a graph with at least two vertices whose vertices w are all *isolated* (that is $deg(w) = 0$) is not connected and hence does not have an Euler circuit. Yet, the degree of each vertex in this graph is even.

8.2 Hamiltonian circuit

In the previous section, we saw a full solution to the Euler circuit problem for undirected graphs. Indeed, with a careful analysis of the problem, we found a simple technique to detect whether a connected graph has an Euler circuit. Moreover, for a graph G that has an Euler circuit, we also presented the *EulerCircuit*(G) algorithm, which outputs an Euler circuit for the graph. One can slightly change the Euler circuit problem as follows. Instead of requiring to use each edge exactly once, one may want to construct a path that visits *each vertex* exactly once. We formally define this as follows.

Definition 8.3. We say that a path $v_0, v_1, \ldots, v_n$ in a graph G is a **Hamiltonian circuit** if *every* vertex of the graph occurs in this path, the last vertex is equal to the first one, and no other vertex is repeated.

Consider for instance examples of graphs presented in Figure 8.2. Which of the graphs have Hamiltonian circuits?

Note that every Hamiltonian circuit is an almost Euler circuit. In particular, no edge in a Hamiltonian circuit is repeated. If we think of vertices as cities, and edges as routes between the cities, then the existence of a Hamiltonian circuit tells us that it is possible to tour through all the cities without ever repeating a city, and returning to the city from which the tour started.

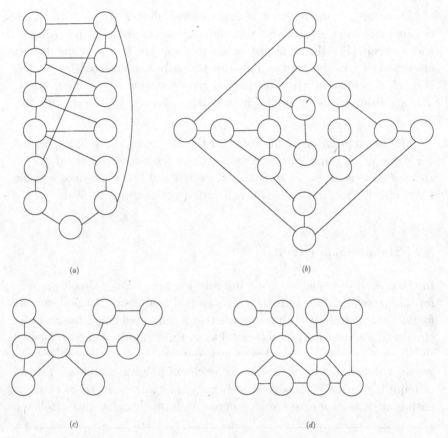

Fig. 8.2: Which of these graphs have Hamiltonian circuits?

Although the concepts of Euler circuits and Hamiltonian circuits sound similar, they are not closely related to each other. For example, there are graphs with Hamiltonian circuits that have no Euler circuits. There are also graphs with Euler circuits but no Hamiltonian circuits. See, for instance, Exercises 3 and 4.

The **Hamiltonian circuit problem** is stated as follows:

Given a graph $G = (V, E)$ determine whether G has a Hamiltonian circuit.

In the rest of this lecture, we discuss this problem and provide one interesting result known as Ore's theorem (proved in 1960). In the previous section, we developed a good understanding of the Euler circuit problem. Namely, we discovered a practical and efficient algorithm that finds Euler circuits for connected graphs. Since the problems sound similar, one may hope that we can also design a practical algorithm that solves the Hamiltonian circuit problem. Interestingly, in contrast to the Euler circuit problem, not much is known about the Hamiltonian circuit problem. As a matter of fact, it is a major open problem whether the Hamiltonian circuit problem can be solved efficiently.

Of course, there is a brute-force algorithm that solves the Hamiltonian circuit problem. Here is its description. A *potential circuit* of length k is any sequence $u_0, \ldots, u_k$ of vertices from V such that (1) k equals the number of vertices of the graph, (2) all the vertices $u_0, \ldots, u_{k-1}$ are distinct, and (3) $u_k = u_0$. Clearly, not every potential circuit is a Hamiltonian circuit. However, given a potential circuit $u_0, \ldots, u_k$ it is easy to determine whether it is indeed a Hamiltonian one. One just needs to check if the conditions (1), (2), and (3) are satisfied.

If a graph G has 2 vertices, then the number of potential circuits is 2. If G has 3 vertices, then the number of potential circuits is 6. If G has 4 vertices, then the number of potential circuits is 24. One can observe that the number of potential circuits in a graph with n vertices is equal to $n!$

The brute-force algorithm proceeds as follows. Examine all potential circuits. For each potential circuit check whether it is a Hamiltonian circuit. If a Hamiltonian circuit is found, then report that a Hamiltonian circuit exists. Otherwise, run through all potential circuits and stop by reporting that a Hamiltonian circuit does not exist.

Although the brute-force algorithm is correct, it is extremely inefficient. In the worst case, it needs to examine *all* potential circuits, which is clearly not feasible for large n.

8.3 Graphs with Hamiltonian circuits

There are some conditions that can be put on graphs that guarantee the existence of Hamiltonian circuits. We single out one of these conditions in the following definition.

Definition 8.4. Let G be a graph with n vertices where $n > 2$. We call the graph G a **club** if for all pairs of distinct vertices u and v not connected by an edge, we have $deg(u) + deg(v) \geq n$.

Assume that we interpret the edge relation between vertices x and y in G as "*x and y know each other*" relation. If no edge exists between x and y, let us interpret this as "*x and y are strangers*". Thus, if a graph G is a club, then for any two strangers in G one of them must know at least half of the vertices of the graph.

It is simple to detect if a given graph G is a club. Indeed, we first run through the adjacency matrix representation of G and output all the strangers (x, y) into a list. Then, for every pair of strangers (x, y), we verify if $deg(x) + deg(y) \geq n$, where n is the number of vertices in G. If so, then G is a club, and otherwise G is not a club. Our next theorem shows that all the graphs that are clubs have Hamiltonian circuits. Note that this theorem is for one direction only; it does not show that a graph with a Hamiltonian circuit must be a club.

Theorem 8.3. *If graph G is a club, then G must have a Hamiltonian circuit.*

Proof.
We prove this theorem by contradiction. So, assume that $G = (V, E)$ is a club but G does not have a Hamiltonian circuit. Let n be the number of all vertices of G.

We can assume that the set of edges, E, is as large as possible in the following sense. If we add a new edge to E, then the resulting graph has a Hamiltonian circuit. We call this the *maximality property* of the edge set E. The reasoning below assumes this maximality property. Note that this assumption requires a proof (see Exercise 10) which we leave to the reader.

Let us add a new edge to E. Since G has the maximality property, the resulting graph must have a Hamiltonian circuit:

$$v_1, v_2, \ldots, v_n, v_1.$$

The circuit must use the new edge. We can assume that the new edge is $\{v_1, v_n\}$. If we remove the new edge from the circuit, then we obtain a path:

$$v_1, v_2, \ldots, v_n.$$

Note that v_1 and v_n are strangers. Since G is a club, we have $deg(v_1) + deg(v_n) \geq n$. We now define two lists of integers, L_1 and L_2, based on the path $v_1, v_2, \ldots, v_n$ as follows:

(1) Put an integer i into the list L_1 if there is an edge between v_1 and v_i.
(2) Put an integer j into the list L_2 if there is an edge between v_n and v_{j-1}.

Clearly, L_1 does not contain 1. The list L_1 also does not contain n since no edge exists between v_1 and v_n. The list L_1 contains exactly $deg(v_1)$ items. Similarly, L_2 does not contain 1 and 2, and has exactly $deg(v_n)$ items.

From our assumption that $deg(v_1) + deg(v_n) \geq n$ and the construction of L_1 and L_2, there must exist at least one integer k that is put in both lists, L_1 and L_2. We leave a proof of this fact to the reader (see Exercise 11). This means, from the definitions of L_1 and L_2, that there exist edges between v_1 and v_k, and between v_{k-1} and v_n.

Thus, we can transform the path v_1, v_2, ..., v_n into a new path by traveling from v_1 to v_{k-1} first, then moving to v_n and traveling from v_n backwards up to v_k, and moving from v_k to v_1. Formally, we produce the following path:

$$v_1, v_2, \ldots, v_{k-1}, v_n, v_{n-1}, v_{n-2}, \ldots, v_k, v_1.$$

This path is clearly a Hamiltonian circuit. Therefore, we have a contradiction to the assumption that G does not have a Hamiltonian circuit. □

8.4 Exercises

(1) Let $G = (V, E)$ be a graph. An **Euler path** in the graph is a path $v_1, v_2, \ldots, v_n$ that traverses each edge of the graph exactly once. Prove that if a connected graph G has a Euler path, then it has exactly zero or two vertices of odd degree.
(2) Determine which graphs in Figure 8.2 have Hamiltonian circuits.
(3) Construct a graph with a Hamiltonian circuit but without an Euler circuit.
(4) Construct a graph with an Euler circuit but without a Hamiltonian circuit.
(5) Let K_n be the graph with $n > 0$ vertices such that there is an edge between every pair of distinct vertices in the graph. As we know these are called *complete graphs* as explained in Section 7.3.

    (a) When does K_n have a Hamiltonian circuit?
    (b) When does K_n have an Euler circuit?

(6) We define the following graphs $K_{n,m}$ called **complete bipartite graphs**.

    (a) All the vertices of $K_{m,n}$ are $-m, -(m-1), \ldots, -1, 1, 2, \ldots, n$. Thus, $K_{n,m}$ has exactly $n + m$ vertices.

    (b) An undirected edge is put between i and j if and only if $1 \le i \le n$ and $-m \le j \le -1$.

Answer the following questions. Explain your reasoning.

    (a) Explain the adjacency matrix representation of $K_{n,m}$.

    (b) When does $K_{n,m}$ have a Hamiltonian circuit?

    (c) When does $K_{n,m}$ have an Euler circuit?

(7) Consider the proof of Theorem 8.2. In the proof, a vertex w is chosen in an almost Euler circuit C_1, such that the degree of w in the remaining graph is not zero. Prove that such a vertex w exists.

(8) Prove that if graph G has a Hamiltonian circuit, then any other graph obtained from G by adding new edges between existing vertices (and preserving the old edges) still has a Hamiltonian circuit.

(9) Let G be a graph with $n > 2$ vertices such that for any vertex v in the graph, we have $deg(v) \ge \frac{n}{2}$. Prove that G has a Hamiltonian circuit.

(10) Consider the proof of Theorem 8.3. At the start of the proof, we assume that E is as large as possible. Explain why we can put this assumption on E.

(11) Consider the proof of the Theorem 8.3. In the proof it is stated that the lists L_1 and L_2 have an integer k in common. Prove this statement.

(12) A **multi-edge graph** consists of vertices and undirected edges, where multiple edges are allowed between vertices. Do the following:

    (a) Define the notion of an Euler circuit for multi-edge graphs.

    (b) Prove that a multi-edge graph has an Euler circuit if and only if the degree of every vertex is even.

(13) Let G be a club with $n > 2$ vertices. Prove, without using Theorem 8.3, that no vertex of degree 0 or 1 exists in G.

(14) Give examples of graphs that have Hamiltonian circuits but that are not clubs.

Programming exercise:

(1) Design and implement an algorithm that decides whether a connected graph has an Euler circuit.

(2) Write a program that, given a connected graph G, outputs an Euler circuit of G.

(3) Implement an algorithm that, given a graph G as input, decides if the graph G is a club.

Lecture 9

Rooted trees

The best time to plant a tree was 20 years ago.
The next best time is now.
Chinese Proverb.

9.1 Basic definitions and examples

Rooted trees are a special type of graphs. They are easier to understand and reason about. They are abstract objects, similar to graphs, used for representing, storing, and manipulating data. In programming practice, in contrast to graphs, rooted trees are easier to process due to their simpler structural properties. Figure 9.1 below shows an example of a rooted tree.

Definition 9.1. A **rooted tree** consists of points and lines between the points. The points are called the **nodes** and the lines are **edges** of the tree. They satisfy the following properties:

(1) There is a special node called the **root** of the tree.
(2) Every node x, apart from the root, is connected to a node y, called the **parent** of x, and x is called a **child** of y.
(3) All nodes, apart from the root, have exactly one parent. The root does not have a parent.
(4) From any node, if we start moving to the parent of the node, from the parent to the parent of the parent, and so on, we eventually reach the root.

In this lecture, we often refer to rooted trees simply as trees. We note, however, in Lecture 17 we use the term tree in a slightly different way.

For the tree in Figure 9.1, the root is 1, and its children are 2 and 3. The parent of nodes 5 and 6 is node 3, and the parent of node 4 is node 2, and so on. When drawing trees, we usually put the root at the top of the tree and parents are always drawn above their children. A node without children is called a **leaf** of the tree and all other nodes are called **interior nodes**. We usually denote trees by the uppercase letter T. This notation assumes that T is defined by its nodes and edges (as graphs are defined by their vertices and edges).

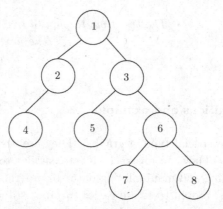

Fig. 9.1: Example of a tree

We can extend the parent-child relationship between the nodes in a tree to an ancestor-descendant relationship as follows. Take a node x_0. Consider a child x_1 of x_0. Consider a child x_2 of x_1. If we continue this on and stop at any point, we produce a **path** $x_0, x_1, \ldots, x_m$ (a path of length m). Note that we do not require x_m to be a leaf. In this path, the node x_0 is an **ancestor** of x_m, and x_m is a **descendant** of x_0. We say that every node is both a descendant and ancestor of itself. From this definition, we see that the root is the ancestor of all the nodes of the tree, and every node is a descendant of the root. In Figure 9.1, for example, nodes 4, 5, 7, 8 are leaves, and 1, 2, 3, 6 are interior nodes. An example of a path is 1, 3, 6, 8. Node 8 is a descendant of 3 but not of 2; and 3 is an ancestor of 7.

In a tree T, take a node x. Consider *all* of its descendants. This determines the following new tree. The root of the new tree is x. The nodes are x and all of its descendants. The edge relation in this new tree is inherited from the original tree. In other words, for x's descendants a and b, the node a is the parent of b in the newly formed tree if and only if a is the parent of b in the original tree T. The new tree is called a **subtree** of T and is denoted by T_x. For example, the subtree T_3 consists of the tree rooted at node 3. It contains nodes 3, 5, 6, 7, and 8. The edges between the nodes are exactly those shown in Figure 9.1.

The **height** of a node x is the length of the longest path from the node to leaves. The height of the tree is the height of its root. In Figure 9.1, the height of the tree is 3. Clearly, the height of the node x is equal to the height of the tree T_x. When we move from the root down to a leaf, along the path, the heights of the nodes decrease one by one. The **level** of x is the length of the path from x to the root. The root is at level 0, its children are at level 1 and so on.

We can put an order on nodes of the tree. One way to do this requires us to have a left-to-right order on the children of every node. For example, in Figure 9.1, the left-to-right order is the natural one. For instance, 2 is left of 3, 5 is left of 6, etc.

Thus, assume that we have a left-to-right order on the children of every node of a tree. We extend this order to *all* nodes of the tree as follows. Take two nodes, say x and y. We say that x **appears before** y in the order if either x is an ancestor of y or the following is true. Follow the paths from x and y to the root; let z be the first node where these two paths meet; these two paths meet from two different children v and w of z, where v is an ancestor of x and w is an ancestor of y; it must then be the case that v is left of w. For example, the nodes of the tree in Fig 9.1, are ordered (according to the left-to-right order) as follows: 1, 2, 4, 3, 5, 6, 7, 8. We give a systematic treatment of ordering the nodes of trees in Section 9.3.

Observe that the existence of the parent-child relationship among the nodes of a tree can be used to define directions on the tree edges. For instance, for an edge $\{x, y\}$, put a direction from x to y if y is a parent of x. In this way, it is clear that every node x of the tree (apart from the root) will have an outgoing edge directed towards the parent of x. One can call this *child-to-parent direction* on edges. The other possible way to define direction on edges is to *"reverse"* the child-to-parent direction that we have just defined. Namely, for an edge $\{a, b\}$ of the tree, put a direction from a to b if a is a parent of b. In this way, any given node a might have several (including 0) outgoing edges towards the children of a.

9.2   Inductive definition of rooted trees

In this section we define rooted trees inductively. This is an alternative definition of rooted trees that uses the concept of iteration. Therefore, this definition is more algorithmic and programmer oriented than the one given in the previous section. Inductive definitions typically explain how to generate the objects of interest. In our case these objects are trees. Understanding inductive definitions is important because one can then reason about these defined objects, as well as their properties. Our inductive definition of trees explains how to generate trees.

The idea is the following. The definition goes in two steps. In the first step we define *small* trees. These are trees that are easiest to comprehend. The first step is called the *basis* of the inductive definition. In the second step, we explain a *method* of building *larger* trees from the smaller ones. The description of the method should be clear and precise. The second step is called the *inductive step* or *induction* of the definition.

Definition 9.2 (Inductive definition of rooted trees). Rooted trees *(or trees, for short) are defined by using the following two rules.*

(1) **Basis***: Any single node x is a* **tree***. This node is the root of the tree.*

(2) **Inductive step***: Assume T_1, ..., T_k are trees with roots x_1, ..., x_k, respectively, such that these trees have no nodes in common. Create a new tree T from T_1, ..., T_k as follows:*

   (a) Create a new node r and declare r to be the root of the tree T.

   (b) Add new edges from r to the nodes x_1, ..., x_k. Thus, each x_i is now a child of r, and we have made r to be the parent of x_1, ..., x_k.

   (c) Declare all nodes of trees T_1, ..., T_k to be nodes of T.

   (d) Declare all edges of trees T_1, ..., T_k to be edges of T.

As an example, we show how this definition can be applied to build the tree in Fig 9.1. The basis of the definition says that the leaf nodes 4, 5, 7 and 8 are trees. We call these trees T_4, T_5, T_7 and T_8, respectively. Apply the second part of the definition to trees T_7 and T_8 to obtain the tree T_6 whose root is 6 and children are 7 and 8. Apply the second part of the definition, again, to trees T_5 and T_6 to obtain the tree T_3 whose root is 3. Similarly, obtain the tree T_2 whose root is 2 and whose leaf is 4. Finally, apply the second part of the definition to trees T_2 and T_3 to obtain the tree T_1, whose root is 1. Thus, T_1 is the tree pictured in Figure 9.1.

Definition 9.2 is a clear example in which objects (in our case, trees) are defined inductively. The basis of the definition explicitly defines the objects that are easiest to comprehend. The second step of the definition gives a clear description of a method that builds larger objects from the objects that have already been built.

9.3 Orders on trees

Given a tree, one may like to process this tree for many different purposes. For example, if the tree represents a file system where the internal nodes represent folders and the leaves represent files, in order to calculate the total disk space used by a folder, one needs to sum up the space used by all its children. Alternatively, perhaps the tree is used to represent the *mother-of* relationship, i.e. node x is the parent of node y if and only if x is y's mother. In this case, one may want to find all the mothers who have exactly two children; this corresponds to counting the total number of nodes with exactly two children nodes. In addition to these, there are many applications where one needs to retrieve information from the data possessed by the nodes. Therefore, it is important to have a systematic way of visiting the nodes in the tree.

We assume the left-to-right ordering is present on children of all the nodes of the tree. Given this, we discuss two common ways for visiting nodes, called $Preorder(T)$ and $Postorder(T)$.

Consider Algorithm 9.1 below. The procedure is called the $Preorder(T)$ algorithm. Given a tree T with root r, the algorithm visits all the nodes of the tree and prints them out.

Algorithm 9.1 $Preorder(T)$ algorithm

---

(1) Print the root r.

(2) Let $x_0, \ldots, x_n$ be all the children of the root r, in left-to-right order.

(3) *While $i \leq n$ do*

    (a) Set T_i be the subtree whose root is x_i.

    (b) Run $Preorder(T_i)$.

    (c) Increment i.

The first line of the algorithm prints the root r. Then, the algorithm considers the children of r. For each child x_i, it calls the $Preorder(T_i)$ algorithm on the subtree rooted at x_i, which prints out x_i and continues by processing the subtree rooted at the first child of x_i. Thus, the algorithm takes the first child x_0 of r and prints x_0. Then it moves to the first child of x_0 and prints that child. Continuing this, the algorithm prints the leftmost leaf in the tree, goes back to the leaf's parent y, and calls the algorithm on the next child of y if it exists. If not, then it moves to the parent z of y, and calls the algorithm on the next child of z if it exists, and so on.

Thus, the $Preorder(T)$ algorithm, using the left-to-right order on the children, prints a node the first time when it visits it. The ordering the nodes of the tree using the $Preorder(r)$ algorithm is called the *preorder*. In Figure 9.2, the order in which the algorithm visits the nodes, according to the preorder relation, is suggested by the arrows. The preorder of the nodes in this tree is: a, b, 0, 3, 4, $+$, 1, c, d, 2, 5, 6, $\star$, 8.

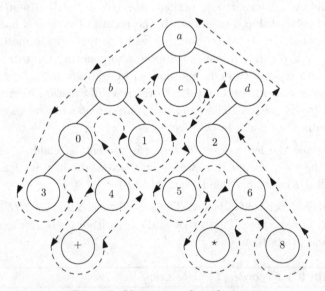

Fig. 9.2: Visiting nodes of a tree

The algorithm below gives us another way of systematically ordering the nodes of a given tree. The order is obtained by printing the nodes the last time when they are visited.

Let us analyze the algorithm. If the root r has no children then the algorithm prints r. Otherwise, the algorithm takes the first child x_0 of r. If

---

Algorithm 9.2 *Postorder(T)* algorithm

---

(1) Let $x_0, \ldots, x_n$ be all the children of the root r, in the left-to-right order.

(2) *While $i \leq n$ do*

  (a) Set T_i be the subtree whose root is x_i.

  (b) Run *Postorder(T_i)*.

  (c) Increment i.

(3) Print the root r.

---

x_0 has no children then the algorithm prints x_0. Otherwise, it moves to the first child of x_0. Continuing this on, the algorithm arrives to the leftmost leaf, say x, and prints the leaf. Afterwards, the algorithm processes all the children of the parent z of x. Once the processing finishes, the algorithm prints z. This continues on until all the nodes of the tree have been visited. The algorithm prints the root r at the end. The ordering of the tree using the *Postorder(r)* algorithm is called the *postorder*. The postorder of the nodes of the tree in Figure 9.2 is 3, $+$, 4, 0, 1, b, c, 5, $\star$, 8, 6, 2, d, a.

9.4 Trees with labels

We organize our directories in an operating system (such as Linux or Windows) into tree-like structures. The nodes of the tree are labeled with file names or directories. Thus, it makes perfect sense to consider trees whose nodes are labeled. These labels represent information or a value associated with the node. It is important to note the distinction between the *names of a node* and the *label of a node*. The main distinction is that, often, we do not want to change the names of the nodes, while the labels of the nodes may change. The labels can be a simple object such as an integer, or a complex object such as a file or text.

Another example of a tree with labels is an expression tree. An expression tree represents an arithmetic expression. To explain the tree representations of arithmetic expressions, we first need to define arithmetic expressions. The definition of an arithmetic expression is an inductive definition, and hence it is similar to the inductive definition for trees. The definition has two steps. First, we define arithmetic expressions called *atomic* arithmetic expressions. These are the easiest expressions to comprehend. Then, we write down rules to build more complicated arithmetic expressions from simpler ones. Here is our definition:

Definition 9.3. Arithmetic expressions are defined by using the following two rules.

(1) **Basis**: Integer variables and integers are arithmetic expressions.
(2) **Inductive step**: Assume e_1 and e_2 are arithmetic expressions. Then, each of the following is also an arithmetic expression: (a) $(e_1 + e_2)$, (b) $(e_1 - e_2)$, and (c) $(e_1 \star e_2)$.

Examples of arithmetic expressions are:

x, 17, -2, $(2+x)$, $((x-y)+(2+x))$, and $((x-y) \star ((x+17)-7))$.

Let us explain why the expression $((x-y) \star ((x+17)-7))$ is an arithmetic expression. Indeed, the expressions x, y, 17 and 7 are arithmetic expressions by the basis of the definition. Apply part (a) to x and 17 to get the expression $(x+17)$, and apply part (b) in the inductive step of the definition to x and y to get $(x-y)$. We now again can apply part (b) to $(x+17)$ and 7 to get $((x+17)-7)$. Finally, apply part (c) to $(x-y)$ and $((x+17)-7)$ to get $((x-y) \star ((x+17)-7))$.

Each arithmetic expression is just a string of symbols. Therefore equality between two arithmetic expressions e_1 and e_2 is defined as equality between the strings e_1 and e_2 (rather than in terms of equality of their values in the domain of integers). For instance, the arithmetic expressions $(x+2)$ and $(2+x)$, and $(x \star y)$ and $(y \star x)$ are *not* equal as strings. Now we define *expression trees* that give labeled tree representation of arithmetic expressions.

Definition 9.4. Let e be an arithmetic expression. The **expression tree** T_e for e is defined by the following two rules:

(1) **Basis**: If e is either an integer variable or an integer, then the expression tree T_e representing e consists of the root only. The root is labeled with e.
(2) **Inductive step**: Assume that e_1 and e_2 are expressions represented by expression trees T_{e_1} and T_{e_2}, respectively. If e is $(e_1 + e_2)$ then the tree T_e with labels, representing $(e_1 + e_2)$, is defined as follows:

(a) Create a new node r and declare r to be the root of the tree T_e. Label r with $+$.

(b) Add edges from r to each of the roots r_1 and r_2 of trees T_{e_1} and T_{e_2}, respectively. Thus, each r_i is now a child of r, and we have made r to be the parent of the roots of the trees T_{e_1} and T_{e_2}.

(c) Set r_1 to be the left child of r, and r_2 to be the right child of r.

(d) Declare all nodes of trees T_1 and T_2 to be nodes of T.

(e) Preserve all the edges of T_{e_1} and T_{e_2}, as well as all the labels.

The expression trees for $(e_1 - e_2)$ and $(e_1 \star e_2)$ are defined similarly. These trees have roots labeled with $-$ and $\star$, respectively, and have subtrees T_{e_1} and T_{e_2}.

Thus, the expression trees are constructed based on the inductive definition of arithmetic expressions. Let us assume that we are given an arithmetic expression e. We want to construct the expression tree, T_e, representing e. This is done by induction in two steps. First, we define expression trees for the easiest arithmetic expressions. Second, we describe a method for building expression trees for more complex arithmetic expressions.

As an example, the expression tree for $((x-y)\star((x+17)-7))$ is shown in Figure 9.3. We comment that in expression trees there is left-to-right-order on children of any given node. Therefore, one can order the nodes of the expression trees (and their labels), using either the preorder or postorder ordering defined in the previous section.

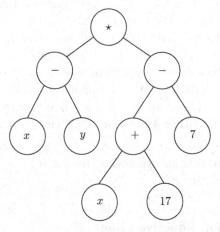

Fig. 9.3: The expression tree for $((x - y) \star ((x + 17) - 7)))$

We also note that in trees representing arithmetic expressions, distinct nodes might have the same labels. For example, the left and the right children of the root in Figure 9.3 are labeled with the symbol $-$. In some textbooks trees with labels are called *labeled trees* and are defined to be such that distinct nodes have distinct labels. In particular, under such labeling, if the tree has n nodes then the number of labels must be equal to n. Our definition of trees with labels does not pose this constraint; namely, we allow distinct nodes of trees to have the same labels.

Now we can prove properties of expression trees. The proofs are based on the inductive definition that builds these trees. As an example, here is one property of expression trees that we will prove.

Proposition 9.1. *If T is an expression tree, then the leaves of T are labeled with either an integer or an integer variable, and the interior nodes are labeled with either $+$, $\star$, or $-$.*

Proof.

The proof is based on the inductive definition that builds expression trees. If tree T is built in the base case, then T has only one node which is its root. Then this root is a leaf in T and it is labeled by either an integer or an integer variable.

Suppose that T is an expression tree for $(e_1 + e_2)$. Consider the expression trees T_{e_1} and T_{e_2} for e_1 and e_2, respectively.

The tree T is built from T_{e_1} and T_{e_2}, and we assume that we have already proved the proposition for the trees T_{e_1} and T_{e_2}. This assumption is called *the inductive hypothesis*. Consider the expression tree T for $(e_1 + e_2)$. The root of this tree must be labeled by $+$. All leaves of T_{e_1} and T_{e_2} are leaves of T. Therefore, by the inductive hypothesis these leaves are labeled by either integers or integer variables. All interior nodes of T_{e_1} and T_{e_2} are also interior nodes of T. Therefore, by the inductive hypothesis these interior nodes are labeled by either $+$ or $\star$ or $-$. Thus, the tree T satisfies the proposition.

The cases when T is an expression tree for $(e_1 - e_2)$ or $(e_1 \star e_2)$ are proved in a similar manner. $\square$

9.5 A schema for inductive proofs

The proof of Proposition 9.1 given in the previous section uses a proof method known as **proof by induction** on trees. We now give an explanation of inductive proofs for trees.

Assume we want to prove a statement S for *all* trees (with labels). The proof of S by induction follows the following three steps.

(1) First, we need to prove that the statement S is true for the case when the tree contains one node only. This is referred to as the *basis* or *base case* of the proof.

(2) Second, we assume that we have a tree T with root r and children x_1, ..., x_k. We consider the subtrees T_1, ..., T_n of T, whose roots are x_1, ..., x_n, respectively. We assume that the statement S has been proved to be true for each of these subtrees. This assumption is called *the inductive assumption*.

(3) Finally, we prove that S is true for T. This part of the proof typically involves using the inductive assumption from the previous step. This step is called *the inductive step*.

If we can do all of the above, then we conclude that the statement S is true for *all* trees. The proof of Proposition 9.1 is a clear example of a proof by induction.

Here is an intuitive explanation of why inductive proofs work. Suppose we have given an inductive proof of some property S for all trees. Say we have a tree T. We want to explain why our inductive proof guarantees that T satisfies S. The tree T is built from smaller trees, say T_1 and T_2. In turn, each of the trees T_1 and T_2 is built from even smaller trees. Since trees become smaller and smaller, we eventually arrive to trees that consist of single nodes only. *The base case* of our inductive proof guarantees that these single node trees all satisfy S. These trees are put together step by step to form the trees T_1 and T_2. Each of these steps preserve the property S. Therefore T_1 and T_2 preserve S. This is what we call *the inductive assumption*. The tree T is build from T_1 and T_2, and our *inductive step* now guarantees that T also satisfies S. This explains why the tree T (and hence all other trees) satisfy S. We will be giving more examples of inductive proofs in later lectures.

9.6 Exercises

(1) Let T be a tree. Prove that there exists exactly one path from any given node to the root of T, when one is only allowed to move from node to parent, and never in the opposite direction.

(2) Consider the tree in Figure 9.3. Order the labels of the nodes of the tree according to the $Preorder(T)$ and $Postorder(T)$ procedures.

(3) Construct expression trees for $((x - y) + ((2 + z) \star y))$, $((((x \star y) + (a + b)) \star z) - 2)$.

(4) Finish the proof of Proposition 9.1 for the cases $(e_1 - e_2)$ and $(e_1 \star e_2)$.

(5) An arithmetic expression e is called **ground** if it contains no integer variables. Each ground arithmetic expression can be evaluated in a natural way. For example, the values of $(2 - 7)$, $((2 + 1) \star (5 - 3))$ are -5 and 6, respectively. Consider an expression tree T for a ground arithmetic expression. The value $val(T)$ of T is defined according to the following rules:

(a) If T consists of only the root then $val(T)$ is the label of the root.

(b) If T is an expression tree for $(e_1 + e_2)$ and T_{e_1} and T_{e_2} are expression trees for ground arithmetic expressions e_1 and e_2 then $val(T) = val(T_{e_1}) + val(T_{e_2})$.

(c) If T is an expression tree for $(e_1 \star e_2)$ and T_{e_1} and T_{e_2} are expression trees for ground arithmetic expressions e_1 and e_2 then $val(T) = val(T_{e_1}) \cdot val(T_{e_2})$.

(d) If T is an expression tree for $(e_1 - e_2)$ and T_{e_1} and T_{e_2} are expression trees for ground arithmetic expressions e_1 and e_2 then $val(T) = val(T_{e_1}) - val(T_{e_2})$.

Prove by induction that for ground expressions e the value of e equals $val(T_e)$, where T_e is the expression tree for e.

(6) A tree is **binary** if every interior node of it has exactly two children. Consider the following algorithm $Inorder(T)$, where T is a binary tree:

(a) Run $Inorder(T)$ on the tree T_l, where l is the left child of the root.

(b) Print out the root.

(c) Run $Inorder(T)$ on the tree T_r, where r is the right child of the root.

Explain what this algorithm does. In particular, apply this algorithm to the tree in Figure 9.3.

(7) Prove by induction that the $Preorder(T)$ and $Postorder(T)$ algorithms both print out all the nodes in the tree T without repetition.

(8) A tree is **finitely branching** if every node of the tree has finitely many children. Give an example of an infinite finitely branching tree.

(9) Prove the following fact known as König's lemma. If a tree is infinite and finitely branching that it has an infinite path. *Hint:* Construct the path by induction.

(10) Define graphs inductively. Your definition should be similar to the inductive definition of rooted trees.

(11) Call a rooted tree **ordered** if there exists a left-to-right-order on the children of every node of the tree. Give an inductive definition of ordered trees.

Programming exercises

(1) Design and implement a *Tree* class of trees with labels, based on its inductive definition.

(2) Add a method to the *Tree* class that prints out the tree so that it is easy to read. You can choose the format for printing the tree.

(3) Write a program that, given a tree, outputs all of its leaves.

(4) Write a program that, given a tree with a left-to-right order on children nodes, outputs the nodes of the tree in

    (a) postorder,

    (b) preorder,

    (c) inorder (see Exercise 6).

(5) Write a program that, given a tree T and a node x, outputs the height of x.

Lecture 10

Sets and operations on sets

The elevator to success is out of order.
You'll have to use the stairs... one step at a time.
Joe Girard.

10.1   Basic definitions and some examples

In this lecture, we learn the basics of sets. We start our discussion with an explanation of sets, their objects, their notation, and ways to create new sets from existing ones. We use all of the notions and notations from this section throughout the forthcoming lectures. Therefore, it is important to read this lecture thoroughly, and to understand it well. Here is an informal definition of a set.

Definition 10.1. A **set** is a collection of objects.

Let us take a set. The objects that form the set may be of any type, such as integers, rational numbers, symbols, programs, edges of a graph, nodes of a tree, and even other sets. The objects that form a given set are called **elements** or **members** of that set. Here are some examples:

(1) The set $\{a, b, c, d\}$ contains four elements: a, b, c, and d.
(2) The set $\{4, 3, 17\}$ contains three elements: 4, 3, and 17.

We describe a set by opening with a left bracket $\{$, listing the elements of the set, and finally closing with a right bracket $\}$. Note that this notation is consistent with our notation for the set of vertices and edges of a given graph (see the first section of Lecture 6).

Given a set and an object, we can ask whether the object is in the set or not. To do this, we use two symbols, $\in$ and $\notin$, which tell us that the object is a member of the set or not, respectively. For example, $a \in \{a, b, c, d\}$, but $17 \notin \{a, b, c, d\}$.

We have already seen examples of sets in our previous lectures. We recall some of them here:

(1) The set $\mathbb{N}$ of all natural numbers. Thus, $\mathbb{N}$ is $\{0, 1, 2, 3, \ldots\}$.
(2) The set $\mathbb{Z}$ of all the integers. Thus, $\mathbb{Z}$ is $\{\ldots, -3, -2, -1, 0, 1, 2, 3, \ldots\}$.
(3) The set $\mathbb{Q}$ of all rational numbers.
(4) For a given graph G we have two sets associated with the graph. The set of its vertices V and the set of its edges E.
(5) The set of all leaves in a given tree.
(6) The set of all divisors of a given integer.
(7) The set of all vertices of a component of an undirected graph.
(8) The set of all paths in a given graph.

Here are more examples of sets that appear in computer science. The set of all programs written in a programming language, the set of all types defined in a program, the set of all values a variable can take, and the set of all current instances of a class in a program.

Given two sets, say A and B, we can compare them. A way to compare them is to look at their members and see how members of one set are related to those of the other. For example, consider the sets $A = \{0, 1, 2\}$ and $B = \{2, 1\}$. Note that all elements of B are elements of A. So, we say that B is a subset of A. Consider $\mathbb{N}$ and $\mathbb{Z}$. All elements of $\mathbb{N}$ are elements of $\mathbb{Z}$. Again, we say that $\mathbb{N}$ is a subset of $\mathbb{Z}$ as we already know from the first lecture. Formally:

Definition 10.2. We say that a set A is a **subset** of a set B, written $A \subseteq B$, if every member of A is also a member of B. If A is a subset of B but B has some members which are not in A, then we say that A is a **proper subset** of B, written $A \subset B$.

From this definition, it is clear that $A \subset B$ always implies $A \subseteq B$.

Consider two sets $A = \{1, 2, 3\}$ and $B = \{3, 1, 2\}$. These two sets are subsets of each other. Namely, $A \subseteq B$ and $B \subseteq A$. However, neither is a proper subset of the other. In fact, both have the same elements. We say that these two sets are equal to each other. Here is a formal definition:

Definition 10.3. We say that sets A and B are **equal** if $A \subseteq B$ and $B \subseteq A$.

Thus, in order to show that two given sets A and B are equal, one needs to show that every member of A is a member of B and that every member of B is a member of A. From the definition of equality between sets, it is clear that the order we list elements of a given set does not change the set. For example, the sets $\{a, b, c\}$, $\{b, a, c\}$, $\{c, b, a\}$ are the same. Also, in listing the elements of a set, we can list each element once without ever repeating it because repeating is redundant. For example, the sets $\{a, b, b, a, a, c, c, c\}$ and $\{a, c, b\}$ are equal.

10.2 Finite and infinite sets

Two important abstract notions in mathematics and computer science are the notions of finiteness and infinity. Finite sets are defined as follows:

Definition 10.4. A set is **finite** if it has *exactly* n elements, for some natural number $n \in \mathbb{N}$. A set is **infinite** if it is not finite.

Thus, a finite set is one which has finitely many elements. For example, the sets $\{2, 3, 4\}$, $\{a, c, d, 10\}$, and $\{3, 2, a, x, y, z, w\}$ are all finite. They have 3, 4, and 7 elements, respectively. A set with no elements is denoted by $\emptyset$ and is called the **empty set**. It is a finite set that has 0 elements. Examples of infinite sets are $\mathbb{N}$, $\mathbb{Z}$ and $\mathbb{Q}$. The set of *all* programs that can be written in a programming language (such as Java) is also an infinite set. The set consisting of all graphs is another example of an infinite set.

There are many ways in which we can describe a set. We discuss two. The first method consists of listing all the elements of a given set. In the previous lectures, we already saw many examples of presenting sets in this manner. These included the sets of vertices and the sets of edges of graphs. Listing elements of a given set usually works for finite sets. However, in order to describe infinite sets or finite sets with many elements, we extend our notation. Since we can not write down *all* the elements of an infinite set, we use the dot notation ... to indicate that the list continues forever. As we have seen, the sets $\mathbb{N}$ and $\mathbb{Z}$ can be written as

$$\{0, 1, 2, 3, 4, 5, 6, 7, 8, 9, \ldots\} \text{ and } \{\ldots, -3, -2, -1, 0, 1, 2, 3, \ldots\},$$

respectively.

The second method describes sets by specifying their elements. This method is referred to as *abstraction*. Say, we have a set A and a property P of elements. We can form a new set consisting of all elements in A that satisfy property P. We write this set as follows:

$$\{x \mid x \in A \text{ and } x \text{ satisfies } P\}.$$

For example, the set $\{n \mid n \text{ is a prime number}\}$ consists of all prime numbers. Similarly, $\{n \mid n = 2k \text{ for some natural number } k\}$ consists of all even natural numbers. Another example is the set $\{x \mid x \text{ is odd and greater than 27}\}$.

10.3　Operations on sets

When we create objects, such as programs, integers, designs, graphs, and so on, we often manipulate them for various purposes. For example, we add, subtract, multiply, and divide integers. We embed Java programs into the body of other programs. We combine multiple graphs to form new graphs. In a file system, we manage our files by deleting, adding, editing, moving and copying files. Another example is in a database. Given the tables in a database, we typically form new tables from existing ones. For instance, we can select several rows in a table, join two tables into one, create a new table that consists of rows that appear in one table but not another, and so on. In general we compare, enumerate, modify, and sort objects we create. All of these are reflections of operations that we can perform on sets.

From a mathematical and programming points of view, operations on sets are methods that create new sets from already existing sets. Here is an interesting example. Take the empty set $\emptyset$. What can we do with it? Well, we can create a new set which is

$$\{\emptyset\}.$$

Now we have two sets: one is $\emptyset$ and the other is $\{\emptyset\}$. We can create another new set from these two sets as follows:

$$\{\emptyset, \{\emptyset\}\}.$$

We then can construct another new set from the ones we have just constructed:

$$\{\emptyset, \{\emptyset\}, \{\emptyset, \{\emptyset\}\}\}.$$

This goes on. What we did is we built new sets from the old ones. This is similar to what we do in programming. We write a program, and can later refer to this program when writing other programs. For instance, at the top of a Java program, it is common for a programmer to import some classes from the Java library, such as the ArrayList class. This class, itself, probably also imports other classes. Achieving a solid understanding of how we manipulate sets can go a long way in helping us understand what goes on with other objects that we may create, use and study (such as programs).

Now we introduce several important operations on sets. For all of these operations, we assume that we are given two sets A and B.

The intersection operation. The **intersection** of sets A and B, written $A \cap B$, is obtained by putting all elements that are members of *both* A *and* B into one single set. One can imagine this as follows. We have a box. For each element a of A, if a belongs to B then we put a into the box. If not, that is $a \notin B$, then we take the next element of A and proceed as before. Thus, the box will hold all elements which are in both A and B. We then name this box with the symbol $A \cap B$. The intersection operation is formally defined as follows:

$A \cap B = \{x \mid x$ is a member of A *and* x is a member of $B\}$.

For example, if $A = \{3, 4, 17, b, c\}$ and $B = \{a, 3, b, c, d\}$, then the intersection $A \cap B = \{b, c, 3\}$.

The union operation. Let us now put all elements of A and all elements of B into one single set. This is called the **union** of A and B. The union of A and B is written as $A \cup B$. Thus, if we pick an element from $A \cup B$ then the element must be either in A or in B. Formally, the union of two sets is defined as follows:

$A \cup B = \{x \mid x$ is a member of A *or* x is a member of $B\}$.

For example, if $A = \{3, 4, 17\}$ and $B = \{a, b, c, 3, d\}$, then the union $A \cup B = \{a, b, c, d, 3, 4, 17\}$.

The set difference operation. The **difference** of A and B, usually written $A \setminus B$, is the set whose elements are those which are members of A but not B.

Example 10.1. Take the set $\mathbb{N}$ of all natural numbers. Let A be the set of all even numbers in $\mathbb{N}$. Then the difference $\mathbb{N} \setminus A$ is the set of all odd natural numbers.

Example 10.2. Take the set $\mathbb{Z}$ of all integers. Let A be the set of all even positive integers. Then the difference $\mathbb{Z} \setminus A$ is the set consisting of all odd natural numbers, 0, and negative integers.

Example 10.3. Take $A = \{3, 4, 17, b, c\}$ and $B = \{a, 3, b, c, d\}$, then the difference $A \setminus B = \{4, 17\}$.

There are some basic properties of these operations on sets. We state them in the theorem below. We prove the first two statements of the theorem and leave the last two as an exercise.

Theorem 10.1. *Let A, B, and C be sets. Then each of the following equalities is true:*

(1) $A \cap A = A$.
(2) $A \cap B = B \cap A$.
(3) $A \cup A = A$.
(4) $A \cup B = B \cup A$.

Proof.
For (1), we need to show that $A \cap A = A$. Recall that two sets are equal if each is a subset of the other. Thus, we need to show that $A \cap A \subseteq A$ and $A \subseteq A \cap A$.

If $x \in A \cap A$ then $x \in A$ and $x \in A$ by the definition of the intersection operation. Hence, $A \cap A \subseteq A$. If $x \in A$ then $x \in A$ and $x \in A$. So $x \in A \cap A$ by the definition of the intersection operation. Hence $A \subseteq A \cap A$. Therefore $A \cap A = A$.

For (2), we need to show that $A \cap B \subseteq B \cap A$ and $B \cap A \subseteq A \cap B$.

If $x \in A \cap B$ then $x \in A$ and $x \in B$ by the definition of the intersection operation. Hence $x \in B$ and $x \in A$. Hence, $x \in B \cap A$ by the definition of the intersection operation. Therefore, $A \cap B \subseteq B \cap A$. Similarly, if $x \in B \cap A$ then $x \in B$ and $x \in A$. Hence $x \in A$ and $x \in B$. Hence, $x \in A \cap B$. Therefore, $B \cap A \subseteq A \cap B$. Thus, $A \cap B = B \cap A$. □

The power set operation. The three operations introduced above produce new sets from existing sets. These are *binary* operations because the arguments to these operations are two sets. There is an interesting operation known as the power set operation. The input to this operation is a single set and the output is also a single set. Thus, the power set operation has one argument and is therefore a *unary* operation. The **power set of** set A is the set of all subsets of A. We write this as $P(A)$. Thus,

$$P(A) = \{X \mid X \subseteq A\}.$$

As an example,

$$P(\{a, b, c\}) = \{\emptyset, \{a\}, \{b\}, \{c\}, \{a, b\}, \{a, c\}, \{b, c\}, \{a, b, c\}\}.$$

This example shows that the power set of a 3 element set has 8 elements. One can easily check that the power set of a 4 element set has 16 elements. It is not hard to derive that if A has exactly n elements, then $P(A)$ has 2^n elements. The empty set $\emptyset$ and the set A are always elements of the power set $P(A)$.

10.4 Exercises

(1) For each of the pair of sets A and B below, determine whether $A \subseteq B$ or $B \subseteq A$ or $A = B$: (a) $A = \{1, 2\}$, $B = \{2, 3, 4\}$; (b) $A = \{a, c\}$, $B = \{c, d, a\}$; (c) $A = \{3, 2, a\}$, $B = \{2, 3, a\}$.

(2) For all sets A and B, prove the equalities $A \cup A = A$ and $A \cup B = B \cup A$.

(3) Let A, B, C be sets. Prove the following equalities:

    (a) $(A \cup B) \cup C = A \cup (B \cup C)$ (*associativity law for the union operation*).

    (b) $(A \cap B) \cap C = A \cap (B \cap C)$ (*associativity law for the intersection operation*).

    (c) $(A \cup B) \cap C = (A \cap C) \cup (B \cap C)$. (*distributivity law*).

    (d) $(A \cap B) \cup C = (A \cup C) \cap (B \cup C)$. (*distributivity law*).

(4) Give examples of sets A and B such that $A \setminus B \neq B \setminus A$.

(5) Let A and B be sets. Prove or disprove the following equalities:

    (a) $P(A \cap B) = P(A) \cap P(B)$.

    (b) $P(A \cup B) = P(A) \cup P(B)$.

(6) Let us fix a set U. For a subset X of U we write $\overline{X}$ to be the set of all $u \in U$ such that $u \notin X$. Thus,

$$\overline{X} = \{u \in U \mid u \notin X\}.$$

We call this set $\overline{X}$ the complement of X. Let X and Y be subsets of U. Prove the following two equalities known as *De-Morgan laws*:

(a) $\overline{X \cup Y} = \overline{X} \cap \overline{Y}$.
(b) $\overline{X \cap Y} = \overline{X} \cup \overline{Y}$.

(7) The **symmetric difference** of two sets A and B, denoted $A \triangle B$, is the following set

$$A \triangle B = (A \setminus B) \cup (B \setminus A)$$

Do the following:

(a) Find $A \triangle B$ for the sets A and B described in Exercise 1.
(b) Prove that for all sets A, B and C if sets $A \triangle B$ and $B \triangle C$ are both finite then $A \triangle C$ is also a finite set.

(8) Consider the set R whose elements are sets X such that X is not a member of X (that is $X \notin X$). Thus,

$$R = \{X \mid X \notin X\}.$$

(a) Prove that if $R \in R$ then $R \notin R$.
(b) Prove that if $R \notin R$ then $R \in R$.

Thus, Definition 10.1 leads us to a contradiction discovered by Bertrand Russell in 1901. This is known as Russell's paradox. This exercise tells us that we need to be careful when we define sets and its elements. Set theory, a branch of mathematics, discusses issues that arise with definitions of sets.

Programming exercises

(1) Design a class Set that can:

(a) support the creation of an empty set or a set consisting of some objects,
(b) compute the union, intersection, set difference, and symmetric difference of sets,
(c) compute the power set of a single set, and
(d) check whether two sets are equal.

(2) Many languages already have at least one class representing sets and implementing the set operations. For example, Java has the HashSet class. Pick one such class, and learn about its underlying representation of sets and how it efficiently implements the set operations. Write a short description of what you have learned.

Lecture 11

Relations on sets

Success is a lousy teacher.
It seduces smart people into thinking they can't lose.
Bill Gates.

11.1 Sequences and Cartesian products

A **sequence** is obtained when we list some objects. For example,

$$4, 1, 3, 2 \quad \text{and} \quad a, c, u, \star, w \quad \text{and} \quad 9, 8, 7, 4, 2, 3, 2, 2, 3$$

are sequences. The length of the first sequence is 4, the second is 5, and the last is 9. So, the **length** of a sequence is the number of elements that occur in that sequence. Note that some elements in the sequence can be repeated. Moreover, the order of the sequence matters, which is in contrast with listing elements of a set. Two sequences are **equal** if they have the same length and the same elements, occuring in the same positions of the sequences. For example, the sequence $1, 2, 5, 4$ is not equal to any of the following sequences: $1, 2, 4, 5$ and $1, 2, 5, 4, 4$ and $1, 2, 5, 4, 0$.

A sequence is **infinite** if its length is not finite. For example, $2, 1, 4, 3, 6, 5, 8, 7, 10, 9, \ldots$ is an infinite sequence. In order to distinguish finite sequences from infinite sequences, finite sequences are called **tuples**. Tuples are usually written within parentheses. For example,

$$(4, 1, 3, 2) \quad \text{and} \quad (a, c, u, \star, w) \quad \text{and} \quad (9, 8, 7, 4, 2, 3, 2, 2, 3)$$

are tuples. A tuple of length 2 is called a **pair** or a 2-**tuple**, a tuple of length 3 is a **triple** or 3-**tuple**, and in general, a tuple of length k is a k-**tuple**.

111

Let A and B be sets. Let us form the set whose elements are pairs of the form (a, b), where $a \in A$ and $b \in B$. For example, for $A = \{1, 2\}$ and $B = \{1, 2, 3\}$, we form the set of pairs

$$\{(1, 1), (1, 2), (1, 3), (2, 1), (2, 2), (2, 3)\}.$$

Thus, from two given sets A and B we produce a new set. The new set obtained is called the Cartesian product of A and B. We give now the following formal definition.

Definition 11.1. The **Cartesian product** of two sets A and B, written $A \times B$, is the set of all pairs (a, b), such that the first element of the pair is a member of A, and the second element of the pair is a member of B. Formally, we write this as $A \times B = \{(a, b) \mid a \in A \text{ and } b \in B\}$. The Cartesian product is also known as the **cross product**.

Any of the sets A and B in the definition can be infinite. For example, when $A = \{1, 2\}$ and $B = \mathbb{N}$, the cross product $A \times B$ is the infinite set of pairs

$$\{(1, 0), (2, 0), (1, 1), (2, 1), (1, 2), (2, 2), (1, 3), (2, 3), (1, 4), (2, 4), \ldots\}.$$

The **Cartesian product** of sets $A_1, \ldots, A_k$, written $A_1 \times \ldots \times A_k$, is the set of all k-tuples $(a_1, \ldots, a_k)$ such that each a_i is a member of A_i, for $i = 1, \ldots, k$. For example, for $A = \{0, 1\}$, $B = \{a, b\}$, and $C = \{(0, 0), x\}$, the set $A \times B \times C$ is the set of 3-tuples

$$A \times B = \{(0, a, (0, 0)), (0, b, (0, 0)), (1, a, (0, 0)), (1, b, (0, 0)), (0, a, x),$$
$$(0, b, x), (1, a, x), (1, b, x)\}.$$

Definition 11.2. If in the cross product $A_1 \times \ldots \times A_k$ all sets A_i are equal to the same set A, then we write the cross product as A^k instead of $A \times \ldots \times A$ where A is repeated k times. Thus, for $k \geq 1$ we have:

$$A^k = \{(a_1, \ldots, a_k) \mid a_1 \in A, \ldots, a_k \in A\}.$$

For instance, for $A = \{x, y\}$ we have:

$$A^3 = \{(x, x, x), (x, y, x), (y, x, x), (y, y, x), (x, x, y), (x, y, y), (y, x, y), (y, y, y)\}.$$

Another example is the set $\mathbb{Z}^2$ that, by definition, consists of all pairs (i,j) of integers. Geometrically, this is the set of all points in the plane whose x and y coordinates are integers. This set represents a two dimensional infinite grid. Similarly, the set $\mathbb{Z}^3 = \{(i,j,k) \mid i,j,k \in \mathbb{Z}\}$ consists of all triples (i,j,k) of integers. Geometrically, this is the set of all points in the three dimensional space whose x, y and z coordinates are integers. This set represents a three dimensional infinite grid.

11.2 Relations

In this section, we define the concept of a relation on a given set. This concept is important in both mathematics and computer science. For instance, in computer science, a relation is the fundamental notion underlying relational databases and their query languages. We start with the definition of a relation:

Definition 11.3. Let A be a nonempty set. A **relation** of arity k on the set A is any subset of A^k. Sometimes relations of arity k are called **k-place relations**. Thus, every k-place relation on A is no more than just a specified set of k-tuples formed from elements of A.

Let R be a k-place relation on set A. When $k = 1$, R is called a **unary relation** on set A. If we write 1-tuples (a) using just the corresponding element, a, then unary relations are simply subsets of A. When $k = 2$, then the relation R is called a **binary** relation. For $k = 3$, the relation R is called a **ternary** relation.

As an example, consider the following set $A = \{0,1,2,3,4\}$. The following are examples of unary relations on A:

(1) $\{0,2,4\}$.
(2) $\{0,4\}$.
(3) $\{1,2,3\}$.

Thus, for instance, the first relation contains elements 0, 2, and 4, and does not contain 1 and 3. We often refer to this as the relation being true on 0, 2, and 4, and being false on 1 and 3.

The following are examples of binary relations on A:

(1) $\{(0,1),(1,2),(2,3),(3,4)\}$.
(2) $\{(0,4),(4,0),(2,4)\}$.
(3) $\{(0,0),(1,1),(2,2),(3,3),(4,4)\}$.

Thus, for instance, the second relation contains pairs $(0,4)$, $(4,0)$, and $(2,4)$, and does not contain all other pairs. Similar to the unary relation case, we often refer to this as the relation being true on pairs $(0,4)$, $(4,0)$, and $(2,4)$, and being false on all other pairs.

The following are examples of ternary relations on A:

(1) $\{(0,1,0),(1,2,1),(2,3,2),(3,4,3)\}$.
(2) $\{(0,0,4),(0,4,0),(1,2,4),(1,1,1)\}$.
(3) $\{(0,0,0),(1,1,1),(2,2,2),(3,3,3),(4,4,4)\}$.

Let A be a set with a k-place relation R on it. We can evaluate the relation R on any given k-tuple $(a_1,\ldots,a_k)$ of elements of A in the following manner. If the k-tuple $(a_1,\ldots,a_k)$ belongs to R, then we say that the value of R *is true on* $(a_1,\ldots,a_k)$. If $(a_1,\ldots,a_k) \notin R$, then we say that the value of R *is false on* $(a_1,\ldots,a_k)$. For example, for $A = \{0,1,2,3,4\}$ and $R = \{(1,4),(4,2),(3,4)\}$, we have the following. The relation R is true on $(1,4)$, $(4,2)$, and $(3,4)$ and is false on all other pairs.

The edges of directed graphs can also be viewed as relations. Indeed, let $G = (V,E)$ be a directed graph. Each edge of the graph is a pair (u,v) where $u \in V$ and $v \in V$. In other words, $E \subseteq V^2$. Therefore, the set of edges E of the graph is an example of a binary relation on the set V of all vertices.

Now we want to count the number of relations on a given finite set. So, let A be a finite set. As we noted above, a unary relation on A is a subset of A. If A has 2 elements, then the number of subsets of A is 2^2. If A has 3 elements, then the number of subsets of A is 2^3. We will later show, or the reader may try to prove this now, that there are 2^n subsets for every set with n elements. Therefore, there are 2^n unary relations on A.

We continue the reasoning above as follows. If a set A has exactly n elements, then the set A^2 of all pairs of A has n^2 elements. Binary relations are subsets of A^2. Therefore, there are 2^{n^2} number of binary relations on A. Similarly, the set of all k-tuples A^k has n^k elements. Therefore, the number of k-place relations on A is equal to 2^{n^k}.

11.3   Binary relations

Many interesting relations on sets are binary relations. Recall that a binary relation on set A is just a subset of A^2, the set of pairs of elements of A. We would like to classify binary relations in some ways. As we mentioned, a good example of a binary relation is the set of all edges of a directed graph (V, E). There are several types of binary relations that often occur in practice and theory. Here we study some of them.

Let A be a set and R be a binary relation on A. Since R is a subset of the set A^2, we can represent R as the following directed graph:

(1) The vertices of the graph are elements of A, and
(2) The edge set of the graph is the set R.

In other words, we put a directed edge from elements a to b if and only if R is true on pair (a, b). We call this a **directed graph representation** of R. It is often convenient to view R in this way. Now, we define several types of binary relations.

Reflexive relations. A binary relation R on A is **reflexive** if $(x, x) \in R$ for *all* $x \in A$. In other words, for R to be reflexive, R must be true on pairs of the form (x, x) for all $x \in A$.

Assume that R is a reflexive relation on A. In the directed graph representation of R, reflexivity is explained as follows. The relation R is reflexive if and only if from *every* vertex $x \in A$, there is a looping edge to x itself.

Now we consider some examples of binary relations and determine if the relations are reflexive.

- On set $\mathbb{Z}$, consider the relation $R = \{(x, y) \mid |x| = |y|\}$. This relation is reflexive because for *every* $x \in Z$, it is the case that $|x| = |x|$, and hence $(x, x) \in R$.
- On set $\mathbb{N}$, consider the relation $Div = \{(x, y) \mid x$ is a factor of $y\}$. This relation is reflexive because for *every* $x \in N$, it is the case that $x = 1 \cdot x$, and hence $(x, x) \in Div$.
- On set $\mathbb{N}$, the binary relation $S = \{(x, y) \mid x + 1 = y\}$ is not a reflexive relation.

For a set A, the **diagonal** of A, denoted by $diag(A)$, is the binary relation $\{(x, x) \mid x \in A\}$. It is called the diagonal because one can view $diag(A)$ as a diagonal across the two-dimensional plane A^2 when A is a one-dimensional line. It is clear that a binary relation R is reflexive if and only if R contains $diag(A)$ as a subset.

We give two more examples from the previous lectures. In Lecture 4, we defined the congruence modulo p relation, denoted by $mod\ p$, on the set $\mathbb{Z}$ of all integers. Recall that an integer n is congruent to an integer m modulo p if $n - m$ is divisible by p. We observed that any integer n is congruent to itself modulo p. Hence the relation

$$mod\ p = \{(n, m) \mid n - m \text{ is divided by } p\}$$

is a reflexive relation on $\mathbb{Z}$. In Lecture 6, we introduced the concept of strong connectivity between the vertices of a given directed graph G. Recall that a vertex x is strongly connected to a vertex y in the digraph G if there exists a path from x to y and there exists a path from y to x. It is an obvious fact that any vertex x in the graph is strongly connected to itself. Hence the relation

$$Connect(G) = \{(x, y) \mid x \text{ and } y \text{ are strongly connected}\}$$

is a reflexive relation.

Symmetric relations. A binary relation R on set A is a **symmetric** relation if for *all* $(x, y) \in R$, it is always the case that $(y, x) \in R$. In the directed graph representation of R, symmetry can be explained as follows. *Whenever* there is an edge from a vertex x to a vertex y, then there must be an edge from y to x. Here are some examples.

- On set $\mathbb{Z}$, consider the relation $R = \{(x, y) \mid |x| = |y|\}$. This relation is symmetric because for all $(x, y) \in R$ it is the case that $|x| = |y|$, and hence $(y, x) \in R$.
- On set $\mathbb{N}$, consider the relation $Div = \{(x, y) \mid x \text{ is a factor of } y\}$. This relation is *not* symmetric because there exists $(x, y) \in Div$ such that $(y, x) \notin Div$. For example, $(1, 5) \in Div$ but $(5, 1) \notin Div$.
- Take any set A. Let R be the empty set. Then, R is symmetric. Indeed, to show that R is symmetric one needs to prove that whenever $(x, y) \in R$ it must be the case that $(y, x) \in R$. Since R is the empty relation, we do not need to check anything.
- Let A be a set. Take any two elements $a, b \in A$. Then, the relation $R = \{(a, b), (b, a)\}$ is a symmetric relation.
- The $mod\ p$ relation on $\mathbb{Z}$ is a symmetric relation as already proved in Lecture 4.

- The relation $Connect(G) = \{((x,y) \mid x \text{ and } y \text{ are connected}\}$ on graph G is a symmetric relation as proved in Lecture 6.

Transitive relations. A binary relation R on set A is a **transitive** relation if whenever $(x,y) \in R$ and $(y,z) \in R$, it must be the case that $(x,z) \in R$. An important point in this definition is this: for R to be transitive it is *not* enough that the transitivity property is satisfied for three particular elements of A; the property *must* be satisfied for *all* triples x, y and z from A. Here are some examples of transitive relations.

- On set $\mathbb{Z}$, consider the relation $R = \{(x,y) \mid |x| = |y|\}$. This relation is transitive because if (x,y) and (y,z) both are in R then $|x| = |y|$ and $|y| = |z|$. Therefore, $|x| = |z|$, which means $(x,z) \in R$.
- On set $\mathbb{N}$, consider the relation $Div = \{(x,y) \mid x \text{ is a factor of } y\}$. This relation is transitive. Indeed, assume that (x,y) and (y,z) are both in Div. Then, x is a factor of y and y is a factor of z. Hence, x is a factor of z. Therefore, $(x,z) \in Div$.
- Take any set A. Let R be the empty set. Then, R is transitive.
- Let A be a set. Take any two elements $a, b \in A$. Then, the relation $R = \{(a,b),(b,a)\}$ is not a transitive relation, because R is true on (a,b) and (b,a), but is not true on (a,a).
- The *mod p* relation on $\mathbb{Z}$ is a transitive relation as we proved in Lecture 4.
- The relation $Connect(G)$ on digraph G is a transitive relation we proved in Lecture 6.

Antisymmetric relations. A binary relation R on set A is **antisymmetric** if *whenever* $(x,y) \in R$ and $(y,x) \in R$, it must be the case that $x = y$. In other words, if R contains a pair (x,y) with $x \neq y$, then R cannot contain the pair (y,x). In the graph representation of R, antisymmetry is explained as follows. For vertices x and y with $x \neq y$ if there is an edge from x to y, then there must not be an edge from y to x. Here are some examples.

- On set $\mathbb{Z}$, consider the relation $R = \{(x,y) \mid x \leq y\}$. This relation is antisymmetric because if $(x,y) \in R$ and $(y,x) \in R$ then that means that $x \leq y$ and $y \leq x$. Therefore, $x = y$.
- On set $\mathbb{N}$, consider the relation $Div = \{(x,y) \mid x \text{ is a factor of } y\}$. This relation is antisymmetric. Indeed, assume that (x,y) and (y,x)

are both in Div. Then x is a factor of y and y is a factor of x. Then, we know that $x \leq y$ and $y \leq x$. This means $x = y$.

- Take any set A. Let R be the empty set. Then R is antisymmetric.
- Let A be a set. Take elements $a_1, b_1, \ldots, a_k, b_k \in A$, all distinct from one another. Then, the relation $R = \{(a_1, b_1), \ldots, (a_k, b_k)\}$ is an antisymmetric relation.
- Let A be a set. For any two distinct elements $a, b \in A$, let relation R be $\{(a, b), (b, a)\}$. The relation R is not an antisymmetric relation, because R is true on (a, b) and (b, a), but $a \neq b$.

11.4   Exercises

(1) Find the cross products $A \times B$, $A \times A$, and $A \times B \times A$ for the following sets: (a) $A = \{1, 2\}$, $B = \{2, 3, 4\}$.   (b) $A = N$, $B = \{1, 2\}$.   (c) $A = \{c, d\}$, $B = \emptyset$.

(2) For sets A, B and C, prove $A \times (B \cup C) = A \times B \cup A \times C$.

(3) Consider the set $A = \{x, y\}$. List down all unary and binary relations on this set. How many ternary relations are there on this set? How many 4-place relations are there on this set?

(4) On the set $A = \{0, 1, 2, 3, 4\}$ consider the following binary relations:

  (a) $R = \{(0, 1), (1, 2), (2, 3), (3, 4)\}$.
  (b) $S = \{(0, 4), (4, 0), (2, 4)\}$.
  (c) $T = \{(0, 0), (1, 1), (2, 2), (3, 3), (4, 4)\}$.

On each of these relations verify reflexivity, symmetry, transitivity, and anti-symmetry properties.

(5) Prove that the *mod p* relation on integers is a reflexive, symmetric and transitive relation.

(6) Prove that the $Connect(G) = \{x, y) \mid x$ and y are connected$\}$ relation defined on graph $G = (V, E)$ is a reflexive, symmetric and transitive relation.

(7) Consider the set $A = \{a, b, c\}$. Do the following:

  (a) Find a relation on A that is symmetric and reflexive but not transitive.
  (b) Find a relation on A that is symmetric and transitive but not reflexive.
  (c) Find a relation on A that is reflexive and transitive but not symmetric.
  (d) Find a relation on A that is not symmetric and not antisymmetric.

(8) Let R be a binary relation on A. Call a relation R' the **minimal symmetric cover** of R if the following are true:

(a) R' is a symmetric relation.
(b) $R \subseteq R'$.
(c) If S is a symmetric relation containing R then $R' \subseteq S$.

Do the following:

(a) On set $A = \{a, b, c, d, e\}$, find the minimal symmetric cover of the relations $R = \{(a, b),\ (c, d),\ (a, e),\ (e, b)\}$ and $S = \{(b, c), (d, c), (c, e), (e, c), (d, d), (b, e), (b, b)\}$.
(b) Design a method that builds the minimal symmetric cover for any given binary relation R.

(9) Let R be a binary relation on A. Call a relation R' the **transitive closure** of R if the following are true:

(a) R' is a transitive relation.
(b) $R \subseteq R'$.
(c) If T is a transitive relation containing R then $R' \subseteq T$.

Do the following:

(a) On set $A = \{a, b, c, d, e\}$, find the transitive closure of the relations $R = \{(a, b),\ (c, d),\ (d, e),\ (e, c), (b, a)\}$ and $S = \{(a, b), (b, c), (c, d), (d, e)\}$.
(b) Design a method that builds the transitive closure for any given binary relation R.

(10) Prove that a relation R on A is both symmetric and antisymmetric if and only if R is a subset of the diagonal, $diag(A)$, of A.

(11) On the power set $P(A)$ define the following relation $\sim$:

$$\sim\ = \{(X, Y) \mid \text{the sets } X \setminus Y \text{ and } Y \setminus X \text{ are both finite}\}.$$

Prove that the relation $\sim$ on $P(A)$ is reflexive, symmetric, and transitive.

Programming exercises

(1) Write a program that as input takes a binary relation on a finite set and determines the following properties of the relation: reflexivity, symmetry, transitivity and antisymmetry.

(2) Write a program that given a finite set A and a binary relation R on A, outputs the minimal symmetric cover of R (see Exercise 8).

(3) Write a program that given a finite set A and a binary relation R on A, outputs the transitive closure of R (See Exercise 9)

Lecture 12

Equivalence relations and partial orders

The mathematical sciences particularly exhibit order, symmetry, and limitation; and these are the greatest forms of the beautiful.

Aristotle.

12.1 Equivalence relations

In this lecture, we continue our study of relations and focus on two special classes of binary relations. These are called equivalence relations and partial orders. We start with equivalence relations, and later we will introduce partial orders. Equivalence relations often appear in many computer science and mathematics applications, especially when one tries to find similarities between objects of interest. The reader should recall the definitions of reflexive, symmetric, transitive and antisymmetric binary relations from the previous lecture.

Definition 12.1. Given a set A, a binary relation E on A is an **equivalence relation** if E is reflexive, symmetric, and transitive. For an equivalence relation E, if we have $(x, y) \in E$ then we say that x and y are **equivalent** (or E-**equivalent**).

Thus, to determine whether a relation E on A is an equivalence relation, one needs to prove the following properties: (1) for all $x \in A$, the pair (x, x) is in E, (2) for all $x, y \in A$, if $(x, y) \in E$, then $(y, x) \in E$, and (3) for all $x, y, z \in A$, if both (x, y) and (y, z) are in E, then $(x, z) \in E$.

In the previous lecture, we proved that the relation *mod p* on integers is reflexive, symmetric, and transitive. Hence, *mod p* is an equivalence relation on $\mathbb{Z}$. Similarly, in the previous lecture for a given digraph $G = (V, E)$, we proved that the relation *Connect(G)* is reflexive, symmetric, and transitive on the set V of all vertices of the digraph G. Hence, *Connect(G)* is an equivalence relation on V.

We now study equivalence relations on sets. The definition below introduces a key concept, the notion of an equivalence class.

Definition 12.2. Let A be a set and E be an equivalence relation on A. For each element $a \in A$, define the following set denoted by $[a]$:

$$[a] = \{x \mid x \in A \text{ and } (x, a) \in E\}.$$

The set $[a]$ is called the **equivalence class** of a.

Example 12.1. As an example, consider the modulo p relation on integers *mod p*. Fix an integer n. The equivalence class of n is the set

$$[n] = \{m \mid m \in \mathbb{Z} \text{ and } n \equiv m \ (mod \ p)\}.$$

We know from Lecture 4 that $[n]$ consists of all integers m such that the integers n and m have the same remainder when divided by p. Therefore, the equivalence classes of the modulo p relation are the congruence classes $[0]$, $[1]$, ..., $[p-1]$ (see Lecture 4).

Example 12.2. Consider the strong connectedness relation, *Connect(G)*, on a directed graph $G = (V, E)$. Fix a vertex v. The equivalence class of v is the set

$$[v] = \{u \mid u \in V \text{ and } (v, u) \in Connect(G)\}.$$

We know from Lecture 6 that $[v]$ consists of all vertices u such that there are paths from v to u, and from u to v. Hence, the equivalence classes of the *Connect(G)* relation are the strongly connected components of the digraph G.

There are two important properties of equivalence relations that we explain below.

Property 1. If E is an equivalence relation on A, then every element of the set A belongs to some equivalence class.

Indeed, take an element $a \in A$. Clearly, $(a, a) \in E$ by reflexivity of E. By the definition of $[a]$, the element a belongs to $[a]$. Hence, a belongs to its own equivalence class $[a]$.

Property 2. Let E be an equivalence relation on A. If $[a]$ and $[b]$ are two equivalence classes, then either $[a] \cap [b] = \emptyset$ (i.e. no overlap) or $[a] = [b]$.

Indeed, if $[a] \cap [b] = \emptyset$, then we are done. Assume that the equivalence classes $[a]$ and $[b]$ have an element c in common. This means that $(a, c) \in E$ and $(b, c) \in E$. Now we want to show that $[a] = [b]$. We begin by showing that $[a] \subseteq [b]$. Indeed, take any element $x \in [a]$. Then, by the definition of the class $[a]$, we have $(a, x) \in E$. We also have $(a, c) \in E$ and $(b, c) \in E$ as given above. Since E is transitive and symmetric, we have $(x, b) \in E$. Therefore $x \in [b]$. Thus, all elements x from $[a]$ are equivalent to b. Therefore, $[a] \subseteq [b]$. Similarly, $[b] \subseteq [a]$. Hence $[a] = [b]$.

Example 12.3. On the set $\mathbb{Z}$ of integers consider the following relation

$$E = \{(x, y) \mid x \cdot y > 0 \text{ or } x = y = 0\}.$$

This is an equivalence relation. Its equivalence classes are:

a) $[1] = \{x \in \mathbb{Z} \mid x > 0\}$,
b) $[-1] = \{m \in \mathbb{Z} \mid m < 0\}$, and
c) $[0] = \{0\}$.

Example 12.4. On the set $\dot{\mathbb{Z}}$ of integers consider the following relation

$$E = \{(x, y) \mid |x| = |y|\},$$

where $|n|$ denotes the absolute value of n. This is an equivalence relation. Its equivalence classes are of the form $\{n, -n\}$, where $n \in \mathbb{Z}$. Thus, there are infinitely many equivalence classes. Note that the equivalence class of 0 is the set $\{0\}$.

12.2 Partitions

The above two properties show that every equivalence relation E on set A determines a collection of subsets that *partition* the set A. These subsets are the equivalence classes and satisfy the following two properties. The first one is that every element of the underlying set A belongs to one of the subsets. The second one is that any two of the subsets are either equal or have no elements in common. We can single out these properties into the following definition.

Definition 12.3. Let A be a set. A collection of non-empty subsets A_1, A_2, A_3, ... of A is called a **partition** of A if the following two properties hold:

(1) $A = A_1 \cup A_2 \cup A_3 \cup \ldots$, and
(2) for all distinct i and j we have $A_i \cap A_j = \emptyset$.

We denote partitions with letter $\mathcal{P}$, possibly with indices.

The first part of the definition states that each element of A belongs to one of the set in $\mathcal{P}$. The second part states that all distinct sets of $\mathcal{P}$ are disjoint with each other.

Example 12.5. An example of a partition of $\mathbb{Z}$ is the following collection $\mathcal{P}$ of subsets:

$$\{0\}, \{-1, 1\}, \{-2, 2\}, \{-3, 3\}, \ldots, \{-n, n\}, \ldots$$

Example 12.6. An example that partitions the set $\mathbb{N}$ is the following collection $\mathcal{P}$ of subsets:

$$\{0, 1\}, \{2, 3\}, \{4, 5, 6, 7\}, \ldots, \{2^n, 2^n + 1 \ldots, 2^{n+1} - 1\}, \ldots.$$

Example 12.7. Let $\mathcal{G}$ be a graph. Partition vertices of V of $\mathcal{G}$ into the following two sets:

$$V_0 = \{v \mid deg(v) \text{ is even } \} \text{ and } V_1 = \{v \mid deg(c) \text{ is odd}\}.$$

Example 12.8. As noted in the paragraph just above the definition, every equivalence relation E on set A determines a partition of A. The partition consists of the equivalence classes of E. Since the partition is dependent on E, we denote it by $\mathcal{P}(E)$. Thus, we can write $\mathcal{P}(E)$ as follows:

$$\mathcal{P}(E) = \{[a] \mid a \in A\}.$$

For instance, as a concrete example, consider the following equivalence relation on $\mathbb{Z}$:

$$E = \{(x, y) \mid x \cdot y > 0 \ \ or \ \ x = y = 0\}.$$

The partition $\mathcal{P}(E)$ consists of the following collection of subsets of $\mathbb{Z}$:

$$\{0\}, \ \{x \in \mathbb{Z} \mid x > 0\}, \ \{m \in \mathbb{Z} \mid m < 0\}$$

The next result below builds equivalence relations from partitions.

Theorem 12.1. *Suppose that we have a partition $\mathcal{P} = \{A_1, A_2, A_3, \ldots\}$ of the set A. Define the following binary relation $E(\mathcal{P})$ on A:*

$$E(\mathcal{P}) = \{(x, y) \mid \text{there is a } j \text{ such that both } x \text{ and } y \text{ belong to } A_j\}.$$

Then the relation $E(\mathcal{P})$ is an equivalence relation.

Proof. The relation $E(\mathcal{P})$ is reflexive because for every x there is an i such that $x \in A_i$. Note that such A_i exists because of the first part of Definition 12.3. Hence, $(x, x) \in E$. The relation E is symmetric because if $x, y \in A_j$ for some j, then $y, x \in A_j$. Transitivity of E is obvious as well but one needs to use the second part of Definition 12.3. $\square$

Let us now take an equivalence class $[a]$ determined by the equivalence relation $E(\mathcal{P})$ constructed in the theorem above. It is easy to see that $[a]$ is equal to the set A_i of the partition $\mathcal{P}$, which contains the element a. Similarly, every A_i is equal to the equivalence class $[a]$ such that $a \in A_i$. Therefore, the partition determined by $E(\mathcal{P})$ is the original partition $\mathcal{P}$.

Thus, we have established a natural correspondence between partitions of the set A and the equivalence relations on A.

Example 12.9. Consider the following partition of $\mathbb{Z}$: $\{0\}$, $\{1, -1\}$, $\{2, -2\}$, $\{3, -3\}$, ... The equivalence relation E associated with this partition is $\{(x, y) \mid x, y \in \mathbb{Z} \text{ and } |x| = |y|\}$.

Example 12.10. Consider the set A of all binary strings over the alphabet $\{0, 1\}$. We define the following partition $A_0, A_1, \ldots$ of the set A, where for each $i \in \mathbb{N}$ we have: $A_i = \{v \mid \text{the length of string } v \text{ is } i\}$. For instance, when $i = 2$ we have $A_2 = \{00, 01, 10, 11\}$. The equivalence relation E associated with this partition is the following:

$$E = \{(v, w) \mid \text{the lengths of } v \text{ and } w \text{ are equal}\}.$$

12.3 Partial orders

Another interesting class of binary relations is the class of partial orders. A partial order arises when we would like to compare objects to one another. For example, the need for comparison might arise when we want to say which object is better than another (for some domain-specific definition of better). The definition of a partial order is quite simple:

Definition 12.4. A binary relation R on set A is called a **partial order** on A if R is reflexive, antisymmetric, and transitive.

There are many natural examples of partial orders. Obvious ones are the natural orders on the set $\mathbb{Z}$ of integers and the set $\mathbb{Q}$ of rational numbers

$$\leq_{\mathbb{Z}} = \{(x,y) \mid x \leq y \text{ and } x,y \in \mathbb{Z}\} \text{ and } \leq_{\mathbb{Q}} = \{(x,y) \mid x \leq y \text{ and } x,y \in \mathbb{Q}\},$$

respectively. Typically, partial orders R on set A are denoted by the symbol $\leq$. Also, instead of writing $(x,y) \in R$, we use the infix notation and write $x \leq y$. Here is another example of a partial order.

Example 12.11. Consider the set A of all binary strings over the alphabet $\{0,1\}$. We say that a string x is a **prefix of a string** y if $y = xv$, for some string v. For example, strings 0, 01, 011, 0110 are prefixes of the string 011011. Consider the relation

$$\leq_{pref} = \{(x,y) \mid x,y \text{ are binary strings and } x \text{ is a prefix of } y\}.$$

This relation is a partial order on A. Indeed, it is reflexive because every string x is a prefix of itself. It is antisymmetric because if x is a prefix of y and y is a prefix of x, then $x = y$. Finally, it is transitive because if x is a prefix of y and y is a prefix of z, then x is clearly a prefix of z.

Example 12.12. For this example, recall that two natural numbers have the same parity if either both are even or both are odd. On the set $\mathbb{N}$, consider the following binary relation $\leq_1$:

> We write $x \leq_1 y$ if and only if either x is odd and y is even
> or x and y have the same parity and $x \leq y$.

We remark the following two obvious properties of $\leq_1$.

(1) For all even numbers a and b we have $a \leq_1 b$ if and only if $a \leq b$.
(2) Similarly, for all odd numbers a and b, we have $a \leq_1 b$ if and only if $a \leq b$.

Now we want to show that $\leq_1$ is a partial order on $\mathbb{N}$. Clearly, $x \leq_1 x$ by the definition of $\leq_1$. Hence $\leq_1$ is reflexive. Let us now prove that $\leq_1$

is a transitive relation. Assume that $x \leq_1 y$ and $y \leq_1 z$. We want to show that $x \leq_1 z$. There are two cases to consider.

Case 1. Assume that x is even. In this case, y must be even because if y were odd then it would be the case that $y \leq_1 x$. With similar reasoning, we infer that z is also even. Therefore, by the remarks above, in this case we have $x \leq_1 z$.

Case 2. Assume that x is odd. Within this case, there are two sub-cases: (A) y is even, or (B) y is odd (and hence $x \leq y$). In case (A), since y is even, it must be the case that z is also even. So, we know that x is odd, and z is even, and thus that $x \leq_1 z$. In case (B), either z is even (in which case, we're once again done) or z is odd and hence $y \leq z$. If the latter, we know that $x \leq y$, $y \leq z$, x is odd and z is odd. It then follows that $x \leq_1 z$.

Finally, to prove that $\leq_1$ is a partial order, we must also show that $\leq_1$ is antisymmetric. We leave this to the reader as an exercise.

The linear order $\leq_1$ defined above on the set $\mathbb{N}$ can be written as follows:

$$1 \leq_1 3 \leq_1 5 \leq_1 \ldots \leq_1 2n + 1 \leq_1 \ldots \leq_1 0 \leq_1 2 \leq_1 4 \leq_1 \ldots \leq_1 2n \leq_1 \ldots.$$

A set A together with a given partial order on it is called a **partially ordered set**. Thus, we can write a partially ordered set as a pair $(A, \leq)$, where $\leq$ is a partial order on A. This is similar with our notation (V, E) for graphs, where by changing E, one changes the graph. In the same manner, in the partially ordered set $(A, \leq)$, we change the partially ordered set by changing the partial order $\leq$.

12.4   Greatest and maximal elements

There are a couple of notions for partially ordered sets that are often used in the study of partial orders. These are the notions of the greatest, maximal, the least and minimal elements.

Definition 12.5. Let $(A, \leq)$ be a partially ordered set.

(1) An element a is called the **greatest element** if $x \leq a$ for *all* $x \in A$.
(2) An element a is called a **maximal element** if there is no $x \in A$ such that $a \leq x$ and $a \neq x$.

Example 12.13. On the set of all negative consider the natural order $\leq$:

$$\ldots \leq -4 \leq -3 \leq -2 \leq -1.$$

The integer -1 is the greats element in this set.

Example 12.14. Consider the partially ordered set

$$(\{0, 00, 01, 11, 000, 010, 011, 110, 111, 0001, 1110, 1111\}, \leq_{pref}),$$

where the $\leq_{pref}$ relation is defined above. Figure 12.1 is a pictorial representation of this partially ordered set. In this representation, an element a is smaller than b with respect to $\leq_{pref}$ if one can move from element a along the lines *upwards* to element b. The elements 0001, 010, 011, 110, 1110 and 1111 are the maximal elements of this partially ordered set.

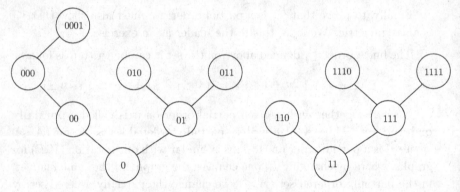

Fig. 12.1: A partially ordered set

As demonstrated in the figure above, in a partially ordered set, a maximal element is one that has no elements above it, whereas the greatest element is one which has all the elements below it. Now let us prove the following simple proposition.

Proposition 12.1. *If $(A, \leq)$ has a greatest element, then the greatest element is maximal. Moreover, in this case, the maximal element is unique.*

Proof.
Indeed, let a be the greatest element. Assume that an element x is such that $a \leq x$. Since a is the greatest, it must be the case that $x \leq a$. Hence, by antisymmetry, we have $a = x$. This shows that a is a maximal element.

Now, let b_1 and b_2 be two maximal elements in the partial order. Then, by the definition, no element x exists such that $x \neq b_1$ and $b_1 \leq x$. However, a is the greatest element. Therefore, $b_1 \leq a$. Hence $b_1 = a$. Similarly, $b_2 = a$. Thus, $b_1 = b_2$. □

Rooted trees, introduced in Lecture 9, can also be considered as partially ordered sets. Indeed, let T be a tree. On the set of nodes of the tree introduce the following relation $\{(x, y) \mid x$ is a descendent of y or $x = y\}$. This relation is a partial order on the set of nodes. Note that the root of the tree is the greatest element of the partial order.

12.5 Hasse diagrams

A convenient way to represent partially ordered sets is through diagrams called *Hasse diagrams*. An example of a Hasse diagram is Figure 12.1 above.

For another example, consider the following partially ordered set $(P(\{a, b, c\}), \subseteq)$. Recall that $P(\{a, b, c\})$ is the power set of $\{a, b, c\}$. So, the elements of $P(\{a, b, c\})$ are $\emptyset$, $\{a\}$, $\{b\}$, $\{c\}$, $\{a, b\}$, $\{a, c\}$, $\{b, c\}$, and $\{a, b, c\}$. The relation $\subseteq$ is a partial order on $P(\{a, b, c\})$. Figure 12.2 gives a pictorial representation of this partial order.

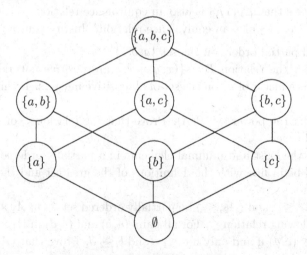

Fig. 12.2: A partially ordered set

In this representation, an element a is smaller than b with respect to $\subseteq$ if one can move from element a along the lines *upwards* to element b.

We now give a more formal explanation of Hasse diagrams.

Assume that we have a partially ordered set $(A, \leq)$. Here we assume that the set A is *finite*. We say that an element $y \in A$ **covers** another element $x \in A$ if $x \leq y$ and no element exists strictly between x and y with respect to the partial order $\leq$. For instance, In Figure 12.1 the element 000 covers 00, the elements 010 and 011 cover 01. Now, to draw a Hasse diagram of $(A, \leq)$ we proceed as follows:

(1) Represent each element of A as a point in the plane.
(2) Draw a line or a curve that goes *upward* from a point x to a point y if y covers x.
(3) The curves might cross each other but must not touch any point apart from their own endpoints.

In this drawing $x \leq y$ if and only if there exists a path going upward from x to y.

12.6   Exercises

(1) Let E_1 and E_2 be equivalence relations on A.

    (a) Prove that $E_1 \cap E_2$ is also an equivalence relation.
    (b) Is $E_1 \cup E_2$ also an equivalence relation? Justify your answer.

(2) List all partial orders on the set $\{a, b, c\}$.
(3) Consider the relation $R = \{(x, y) \mid x, y$ are positive natural numbers and x is a factor of $y\}$ on the set of all positive integers. Is this a partial order?
(4) Consider the power set $P(A)$. Prove that the relation $\subseteq$ on $P(A)$ is a partial order.
(5) Define the least and minimal elements in a partially ordered set. This should be in line with the definitions of the greatest and the maximal elements.
(6) Let $(A_1, \leq_1)$ and $(A_2, \leq_2)$ be partially ordered sets. On $A_1 \times A_2$ define the following relation $\leq$. For all pairs (a, b) and (c, d) in $A_1 \times A_2$ we set $(a, b) \leq (c, d)$ if and only if $a \leq_1 c$ and $b \leq_2 d$. Prove that $(A_1 \times A_2, \leq)$ is a partially ordered set.
(7) Draw Hasse diagrams of the following partially ordered sets:

    (a) $A = \{1, 2, 3, 4, 5, 6, 10, 12, 15, 20, 30, 60\}$, and for all $x, y \in A$ we have $x \leq y$ if and only if x is a factor of y.

(b) $A = \{x \mid x$ is a binary string of length at most 3$\}$, and for all $x, y \in A$ we have $x \leq y$ if and only if x is a prefix of y.

(c) $A = \{0, 1, 2, 3, 4\} \times \{0, 1, 2, 3, 4\}$, and for all pairs $(a, b), (c, d) \in A \times A$ we have $(a, b) \leq' (c, d)$ if and only if $a \leq c$ and $b \leq d$.

(8) Consider the following set $A = \{2, 3, 4, 5, 6, 7, 8, 9, 11, 12, 13, 14, 15\}$. On this set, consider the relation $R = \{(x, y) \mid x$ is a factor of $y\}$. Do the following:

    (a) List all maximal and minimal elements of this partially ordered set.

    (b) Draw a Hasse diagram of this partially ordered set.

(9) On the set $\mathbb{N}$, consider the following binary relation $R = \{(x, y) \mid$ there is a natural number n such that $y = 2^n \cdot x\}$. Do the following:

    (a) Prove that this is a partial order of $\mathbb{N}$.

    (b) What are the minimal elements of the partially ordered set $(\mathbb{N}, R)$?

    (c) Draw a Hasse diagram of this partially ordered set.

Programming exercises

(1) Write a program that on input (A, E) determines if E is an equivalence relation on A.

(2) Write a program that given an equivalence relation E on A, outputs all equivalence classes of E.

(3) Write a program that given a partially ordered set $(A, \leq)$:

    (a) Determines if $(A, \leq)$ has a greatest element and a least element.

    (b) Outputs all maximal and minimal elements of the partial order.

    (c) Draws a Hasse diagram representing the partial order.

Lecture 13

Databases and relational structures

Information is the oxygen of the modern age.
It seeps through the walls topped by barbed wire,
it wafts across the electrified borders.

Ronald Reagan.

13.1 A toy example of a database

In this section, we establish a connection between databases and relations. We discuss a toy example of a database that contains information about the employees of a small car repair company. The company consists of ten employees, and the database provides some information about these people. This information is presented using several tables (see Figure 13.1).

We now describe these tables in detail. The *Employees* table associates the *IDs* of the employees with their names. The next table *BankAccounts* contains information about the bank account details of employees for salary payment purposes. It lists the bank and bank account number for each employee. To boost company morale, the car repair company celebrates each employee's birthday, and thus maintains a table of the employee's birth dates in the table *BirthDates*. The next table, *DirectlyReports*, contains information about the management structure of the company. If there is a row (X, Y) in the *DirectlyReports* table, then this means that X reports to Y. For example, the employees with IDs 01, 02 and 03 directly report to the employee with ID 00. Finally, *AACertified* lists all employees who have passed the Automobile Association certification test and are thus eligible for higher wages.

Each of these tables consists of columns and rows. Each column has

Employees:

ID	FName	LName
00	Elaine	Aziz
01	Robert	Simmons
02	Andrei	Goncharov
03	Michael	Love
04	Steven	Hosking
05	Nevil	Hosking
06	Bruce	Fong
07	Marina	McDonald
08	Ann	Roberts
09	Cynthia	Mendes

BankAccounts:

EID	Bank	AcctNum
00	National	712456001
01	Union	713456501
02	National	712456001
03	West-Union	512566001
04	West-Union	712406006
05	Lake Bank	212456708
06	Lake Bank	415456051
07	National	912454087
08	National	032456548
09	Lake Bank	945456100

BirthDates:

EID	Y	M	D
00	1960	Jan	30
01	1971	Jan	10
02	1974	Mar	11
03	1980	Dec	22
04	1964	Aug	23
05	1964	Sep	30
06	1968	Sep	1
07	1970	Jun	15
08	1981	Jun	3
09	1982	Nov	16

DirectlyReports:

EID	MgrID
01	00
02	00
03	00
04	01
05	02
06	02
07	03
08	03
09	08

AACertified:

EID
00
06
08
09

Fig. 13.1: Example database for a car repair company

a name. For example, the *BirthDates* table has four columns named and listed as follows: *EID*, *Y*, *M*, *D*. These names of columns are called the *attributes* of the table. For example, *AACertified* table has one attribute and *BankAccounts* table has three attributes.

Clearly, our example database has five tables. The collection of all entries in these five tables is called the **domain** of the database. Thus, the domain D of our database consists of the names of people, their IDs, their account numbers, the names of banks, years, months, and days. Each table of the database describes a relationship between elements in the domain. Each row in a table describes a fact. For example, the row $(00, 1960, January, 30)$ in *BirthDates* table tells us that Elaine Aziz was born on January 30 in 1960.

Consider the domain D of the database. Take a table T in the database. The most important part of T is its rows because they contain all the information. Each row is a tuple. No two rows in T are the same. Moreover, if we change the order of rows and thus construct a new table T', then the tables T and T' convey the same information. The order of the rows in the table, therefore, is not important. All the rows of T have the *same* length. This length is called the **arity** of T. The arity of the table T is thus the number of its attributes. For example, the arity of the table $AACertified$ is 1. The length of rows in the table $DirectlyReports$ is 2, and thus this table is of arity 2. The arity of the table $Birthdates$ is 4.

Each table T is a set of tuples of the same length. Each tuple is formed from the elements of the domain D. A table T of arity k is thus a subset of the set D^k, where D^k is the set of all k-tuples from D. Hence, each table of arity k is a k-place relation on D. For instance, the table $DirectlyReports$ can be written as:

$$\{(01,00),(02,00),(03,00),(04,01),(05,02),(06,02),(07,03),(08,03),(09,08)\}.$$

Thus, the table $DirectlyReports$ is clearly a binary relation on the domain set D. Similarly, $AACertified$ table is the following unary relation $\{00,06,08,09\}$ on set D. Thus, $AACertified$ is a unary relation, the $BankAccounts$ and the $Employees$ tables are both ternary relations, and the $Birthdates$ table is a relation of arity 4 on the set D.

13.2 Relational structures

Each relation on a domain D can be viewed as a table. Indeed, if we are given a k-ary relation R on D, then R can be viewed as a table in which each row has length k. The table is built in such a way that a k-tuple $(d_1, \ldots, d_k)$ is a row of the table if and only if R is true on $(d_1, \ldots, d_k)$. This simple observation leads us to the following mathematical definition of a relational database:

Definition 13.1. A **relational structure** is a tuple $(D; R_1, \ldots, R_n)$, where

(1) D is a nonempty set called the **domain**,
(2) All $R_1, R_2, \ldots, R_n$ are relations on the domain D called **basic** relations.

In computer science, it is often the case that relational databases are identified with relational structures. This identification allows one to employ purely mathematical tools to study relational databases, their properties, and build new relations from the ones already given. We give two more examples of relational structures.

Example 13.1. The relational structure $(D; R_1, R_2, U)$ is defined as follows. The domain D is $\{0, 1, 2, 3, 4\}$. The binary relations R_1 and R_2 are $\{(0, 4), (4, 0), (0, 1), (0, 2), (0, 3), (4, 1), (4, 2), (4, 3)\}$ and $\{(0, 1), (0, 2), (0, 3), (0, 4), (2, 3)\}$, respectively. The unary relation U is $\{0, 2, 4\}$.

We can represent this relational structure using tables as shown in Figure 13.2.

R_1:

Col 1	Col 2
0	4
4	0
0	1
0	2
0	3
4	1
4	2
4	3

R_2:

Col 1	Col 2
0	1
0	2
0	3
0	4
2	3

U:

Col 1
0
2
4

Fig. 13.2: Example of binary relations R_1, R_2 and unary relation U.

Example 13.2.

(1) The domain D is the following set: $D = \{v \mid v$ is a non-empty binary string over the alphabet $\{0, 1\}$ of length at most 3$\}$.
(2) Here are the relations:

   (a) $R_1 = \{(x, y) \mid x$ is a prefix of $y\}$.
   (b) $R_2 = \{(x, y) \mid$ the length of x equals to the length of $y\}$.
   (c) $R_3 = \{(x, y, z) \mid (y = x0$ and $z = y0)$ or $(|y| = 2$ and $x = 00$ and $z = 111\}$, where $|y|$ refers to the length of string y.
   (d) $R_4 = \{(x, y) \mid y = x0\}$.
   (e) $R_5 = \{(x, y) \mid y = x1\}$.

As above, we can now represent these relations as tables. For example, a table representation of relation R_3 is given in Figure 13.3.

x	y	z
0	00	000
1	10	100
00	00	111
00	01	111
00	10	111
00	11	111

Fig. 13.3: Relation R_3

Finally, it is worth noting that any graph $G = (V, E)$ can also be regarded as a relational structure. Indeed, for a graph $G = (V, E)$ we can treat the set V of vertices as the domain and the set E of edges as the binary relation, represented as a table of arity 2. We also note that our definition of relational structures does not prohibit the domain D from being infinite.

13.3   Exercises

(1) Consider the database described in the first section. Write down the relations and their arities that are represented by the tables *Employees*, *AACertified*, and *BankAccounts*.

(2) Consider the graph G representing a binary relation R on domain D. How do we change G to represent the relation $D^2 \setminus R$?

(3) Write down tables for each of the relations in the following relational structure

$$(D; S, Add, Mult, Div, \leq), \text{ where}$$

- The domain D is $\{0, 1, 2, 3, 4\}$.
- The successor relation: $S = \{(x, y) \mid y = x + 1 \text{ and } x, y \in D\}$.
- The addition relation: $Add = \{(x, y, z) \mid x + y = z \text{ and } x, y, z \in D\}$.
- The multiplication relation: $Mult = \{(x, y, z) \mid x \cdot y = z \text{ and } x, y, z \in D\}$.
- The order relation: $\leq = \{(x, y) \mid x \leq y \text{ and } x, y \in D\}$.

(4) Assume that a relation R consists of exactly n tuples. How many tables are there that represent the relation?

(5) Suppose that we have two relations R_1 and R_2 of the same arity.

    (a) Describe a method that builds a table for the union $R_1 \cup R_2$ of the relations.

    (b) Describe a method that builds a table for the intersection $R_1 \cap R_2$ of the relations.

Lecture 14

Relational calculus

Information is not knowledge.
Albert Einstein.

The purpose of this lecture is to present methods for building new relations from existing ones. From a relational databases point of view these methods build new tables from the existing tables. Below, we recall the example of a relational structure from the previous lecture.

Example 14.1. The relational structure $(D; R_1, R_2, U)$ is defined in the following manner:

(1) The domain $D = \{0, 1, 2, 3, 4\}$,
(2) The relation $R_1 = \{(0, 4), (4, 0), (0, 1), (0, 2), (0, 3), (4, 1), (4, 2), (4, 3)\}$,
(3) The relation $R_2 = \{(0, 1), (0, 2), (0, 3), (0, 4), (2, 3)\}$, and
(4) The unary relation $U = \{0, 2, 4\}$.

In Lecture 13, we represented this relational structure by writing down the tables for relations R_1 and R_2. Another nice representation for viewing binary relations, such as R_1 and R_2, is to use directed graphs, as given in Figure 14.1. We will use this example in explaining the methods.

Assume that we are given a non-empty, and possibly infinite set A. We introduce the following operations on relations over the set A. These operations are called the **first order logic operations**. One of the main applications of these operations is that they form the mathematical foundation of many database query languages such as SQL, Datalog, and XQuery.

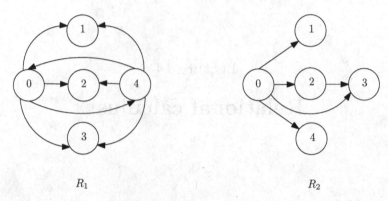

R_1 R_2

Fig. 14.1: Directed graph representations of R_1 and R_2

14.1   Boolean operations

Since relations are just sets we can use the the set-theoretic operations of union, intersection and complementation to form new relations from old ones. We briefly recast them again below.

The union operation. This operation takes as input two relations of the same arity and outputs their union. Formally, given two relations, R_1 and R_2, both of arity k, their union is the relation $R_1 \cup R_2$. Thus, a tuple $(a_1, \ldots, a_k)$ belongs to $R_1 \cup R_2$ if and only if either $(a_1, \ldots, a_k) \in R_1$ *or* $(a_1, \ldots, a_k) \in R_2$.

The union operation applied to R_1 and R_2 from Example 14.1 produces $R_1 \cup R_2 = \{(0, 4), (4, 0), (0, 1), (0, 2), (0, 3), (4, 1), (4, 2), (4, 3), (2, 3)\}$. The relation $R_1 \cup R_2$ is thus true on (x, y) if and only if either R_1 is true on (x, y) or R_2 is true on (x, y).

The intersection operation. This operation takes as input two relations of the same arity and outputs their intersection. Formally, given two relations, R_1 and R_2, both of arity k, their intersection is the relation $R_1 \cap R_2$. Thus, a tuple $(a_1, \ldots, a_k)$ belongs to $R_1 \cap R_2$ if and only if $(a_1, \ldots, a_k) \in R_1$ *and* $(a_1, \ldots, a_k) \in R_2$.

For instance, the intersection $R_1 \cap R_2$ of R_1 and R_2 from Example 14.1 is the binary relation $\{(0, 1), (0, 2), (0, 3), (0, 4)\}$. The relation $R_1 \cap R_2$ is thus true on all pairs (x, y) such that both R_1 and R_2 are true on (x, y).

The complement operation. This operation takes as input a relation and outputs its complement. Formally, given a relation R of arity k, its complement $\overline{R}$ is the relation $A^k \setminus R$. Thus, a tuple $(a_1, \ldots, a_k)$ belongs to

$\overline{R}$ if and only if $(a_1, \ldots, a_k) \notin R$.

In Example 14.1, the complement of U is $\{1, 3\}$. The complement of R_1 is the relation $\overline{R_1}$ that is true on (x, y) if and only if R_1 is false on (x, y). For example, $\overline{R_1}$ is true on $(1, 2)$ and $(2, 2)$. The relation $\overline{R_1}$ has exactly 17 pairs.

14.2 The existentiation operation

The existentiation operation is also called the **projection operation**. This operation takes as input a relation R of arity k and an i such that $1 \le i \le k$. If $k = 1$ (that is, R is a unary relation), then this operation outputs *True* in case $R \neq \emptyset$, and outputs *False* in case $R = \emptyset$. If $k > 1$, then the operation outputs the following $(k-1)$-place relation, denoted by $\exists x_i R$, called an **existentiation of R**:

$$\{(a_1, \ldots, a_{i-1}, a_{i+1}, \ldots, a_k) \mid \text{there is an } a \in A \text{ such that}$$
$$(a_1, \ldots, a_{i-1}, a, a_{i+1}, \ldots, a_k) \in R\}.$$

Thus, for the relation $\exists x_i R$ to be true on $(b_1, \ldots, b_{k-1})$, it must be the case that the relation R is true on $(b_1, \ldots, b_{i-1}, a, b_i, \ldots, b_{k-1})$ for *some* $a \in A$.

Consider Example 14.1. Both $\exists x_1 U$ and $\exists x_1 \overline{U}$ have values *True* since both U and $\overline{U}$ are not empty. For relation R_1, we have the following:

$$\exists x_1 R_1 = \{0, 1, 2, 3, 4\} \quad \text{and} \quad \exists x_2 R_1 = \{0, 4\}.$$

Similarly, for relation R_2, we have the following:

$$\exists x_1 R_2 = \{1, 2, 3, 4\} \text{ and } \exists x_2 R_2 = \{0, 2\}.$$

These are now unary relations.

There are two ways to think about the existentiation operation. The first is that the operation can be viewed as a *search*. Assume that we are given a binary relation R, on some ordered set A whose elements are ordered as follows:

$$a_0, a_1, a_2, a_3, \ldots.$$

Consider the unary relation $\exists x_1 R$. Suppose we want to see if an element $b \in A$ is in $\exists x_1 R$. Algorithmically, this can be explained as follows. On input $b \in A$, execute the following instructions:

(1) Initialize variable i to 0.
(2) If $(a_i, b) \in R$, then output b, and stop.

(3) Increment i, and repeat line (2).

If b is in $\exists x_1 R$, then, by definition, there exists an $a_m \in A$ such that $(a_m, b) \in R$. Hence, the instructions above output b in at most m iterations of line 2) of the procedure above. The converse is also true. If the above instructions output b, then this means that an element a_m is found such that $(a_m, b) \in R$. The element a_m, for which $(a_m, b) \in R$, is usually referred to as a *witness* for b to be in $\exists x_1 R$. Thus, the instructions above search for a witness a_m to be found such that $(a_m, b) \in R$. If b is not in $\exists x_1 R$, then the instructions above run until all elements of A have been processed. Note that if A is an infinite set then the process above runs *forever* searching for a witness.

The second, perhaps, more instructive way to view the existential operation is as follows. Assume that the k-place relation R is represented as a table T. So, each row in T has length k and $k > 1$. Build the following new table defined below:

- **Contracting:** Remove the ith column from the table T. Let T_1 be the new table.
- **Cleaning:** For every row j of T_1, remove the row if the row is identical to some row above it. Output the resulting table and call it T_2.

The table T_2 represents the relation $\exists x_i R$.

The introduced operations can be used to build new relations from given ones. For instance, for relations R_1 and R_2 from Example 14.1, here are some new relations that we can build:

$$R_1 \cup R_2, \ R_1 \cap \overline{R_2}, \ \exists x_1 R_1, \ \exists x_2 R_1, \ \overline{(R_1 \cap \overline{R_2})}, \ \exists x_2 (R_1 \cup R_2),$$
$$\exists x_2 (\overline{(R_1 \cap R_2)}), \ \exists x_1 \overline{R_2}, \text{ etc.}$$

The $\forall$-operation. Often, this operation is called the **universal projection** operation. This operation takes as input a relation R of arity k and an i such that $1 \leq i \leq k$. If $k = 1$ (that is, R is a unary relation), then this operation outputs the value *True* in case $R = A$, and outputs *False* in case $R \neq A$. If $k > 1$ then the operation outputs the relation $\forall x_i R$ of arity $k - 1$ defined as follows

$$\{(a_1, \ldots, a_{i-1}, a_{i+1}, \ldots, a_k) \mid \text{for all } a \in A, \text{ we have}$$
$$(a_1, \ldots, a_{i-1}, a, a_{i+1}, \ldots, a_k) \in R\}.$$

Thus, for the relation $\forall x_i R$ to be true on $(b_1, \ldots, b_{k-1})$, it must be the case that the relation R is true on $(b_1, \ldots, b_{i-1}, a, b_i, \ldots, b_{k-1})$ for *all* $a \in A$.

Consider Example 14.1. Then, both $\forall x_1 U$ and $\forall x_1 \bar{U}$ have values *False*.

For relation R_1 in Example 14.1, we have the following: $\forall x_1 R_1 = \emptyset$ and $\forall x_2 R_1 = \emptyset$. Indeed, for each element $x \in \{0, 1, 2, 3, 4\}$ in the domain of the database, it is the case that $(x, x) \notin R_1$. Therefore, x does not belong to $\forall x_1 R_1$, by the definition of the universal projection operation.

Consider the relation $S_1 = \{(x_1, x_2) \mid x_1 = x_2\} \cup R_1$. In other words, we slightly changed the relation R_1 by adding to it all the pairs of the form (x, x), where x is in the domain of the database. This is a binary relation and $\forall x_2 S_1 = \{0, 4\}$. Indeed, for relation S_1 and the elements 0 and 4 we have all $(0, 0), (0, 1), (0, 2), (0, 3), (0, 4)$ are in S_1 and all $(4, 0), (4, 1), (4, 2), (4, 3), (4, 4)$ are in S_1. Therefore, by the definition of the universal projection, we have $0, 4 \in \forall x_1 S_1$.

Similarly, for relation R_2, we have the following: $\forall x_2 R_2 = \emptyset$. Let's form another binary relation

$$S_2 = \{(x_1, x_2) \mid x_1 = x_2\} \cup R_2.$$

Again, one can check that $\forall x_2 S_2 = \{0\}$. These S_1 and S_2 are now unary relations.

The following proposition establishes the relationship between existential and universal operations:

Proposition 14.1. *Let R be a k-place relation with $k > 1$. Let i be such that $1 \leq i \leq k$. Then the following equality is true:*

$$\exists x_i R = \overline{\forall x_i \, \overline{R}}.$$

Proof.
We need to show that following two relationships hold:

$$\exists x_i R \subseteq \overline{\forall x_i \, \overline{R}} \quad \text{and} \quad \overline{\forall x_i \, \overline{R}} \subseteq \exists x_i R.$$

Indeed, assume that $(a_1, \ldots, a_{i-1}, a_{i+1}, \ldots, a_i) \in \exists x_i R$. Then, there is a b such that

$$(a_1, \ldots, a_{i-1}, b, a_{i+1}, \ldots, a_n) \in R.$$

Hence, it is not true that for all $x \in A$ we have

$$(a_1, \ldots, a_{i-1}, x, a_{i+1}, \ldots, a_n) \notin R.$$

Therefore, we have $(a_1, \ldots, a_{i-1}, a_{i+1}, \ldots, a_n) \notin \forall x_i \overline{R}$. Hence, $(a_1, \ldots, a_{i-1}, a_{i+1}, \ldots, a_n) \in \overline{\forall x_i \overline{R}}$.

Similarly, assume that $(a_1, \ldots, a_{i-1}, a_{i+1}, \ldots, a_n) \in \overline{\forall x_i \overline{R}}$. Then, it is not true that for all $x \in A$ we have $(a_1, \ldots, a_{i-1}, x, a_{i+1}, \ldots, a_n) \notin R$. This means there must exists an element $b \in A$, such that

$$(a_1, \ldots, a_{i-1}, b, a_{i+1}, \ldots, a_n) \in R.$$

From the definition of the projection operation, we have $(a_1, \ldots, a_{i-1}, a_{i+1}, \ldots, a_i) \in \exists x_i R$. □

14.3 Other operations

The cylindrification operation. This operation takes as input a relation of arity k and outputs a relation of arity $k + 1$ as follows. Given a relation R of arity k, its cylindrification denoted by $c(R)$, is the set $\{(a_1, \ldots, a_k, a) \mid (a_1, \ldots, a_k) \in R$ and $a \in A\}$. Note that the cylindrification operation increases the arity of an input relation by 1. It is also clear that $c(R)$ can equivalently be defined as the Cartesian product $R \times A$.

For example, the cylindrification $c(U)$ of U in Example 14.1 produces the binary relation: $\{(0,0), (0,1), (0,2), (0,3), (0,4), (2,0), (2,1), (2,2), (2,3), (2,4), (4,0), (4,1), (4,2), (4,3), (4,4)\}$.   The cylindrifications of R_1 and R_2 are ternary relations. The cylindrification of $\overline{U}$, the complement of the unary relation U, is

$$c(\overline{U}) = \{(1,0), (1,1), (1,2), (1,3), (1,4), (3,0), (3,1), (3,2), (3,3), (3,4)\}.$$

The instantiation operation. This operation takes as input a relation R of arity k, an element $a \in A$, and an i such that $1 \leq i \leq k$. If $k = 1$, then this operation outputs the value *True* or the value *False* as follows. If R is true on a, then the output is *True*, and otherwise the output is *False*. Now, assume that $k > 1$. Define the relation $Inst(R, a, i)$, of arity $k - 1$, as follows

$$\{(a_1, \ldots, a_{i-1}, a_{i+1}, \ldots, a_k) \mid (a_1, \ldots, a_{i-1}, a, a_{i+1}, \ldots, a_k) \in R\}.$$

For relations R_1 and R_2 in Example 14.1, we have $Inst(R_1, 0, 1) = \{1, 2, 3, 4\}$, $Inst(R_1, 3, 2) = \{0, 4\}$, $Inst(R_2, 2, 1) = \{3\}$, and $Inst(R_2, 3, 1) = \emptyset$.

The rearrangement operation. First, we need a simple concept of a permutation. A **permutation** of the sequence $1, \ldots, k$ is any re-ordering

$i_1, i_2, \ldots, i_k$ of it. For example, there are 6 permutations of the sequence $1, 2, 3$ which are $1, 2, 3$, $1, 3, 2$, $2, 1, 3$, $2, 3, 1$, $3, 1, 2$, and $3, 2, 1$. It can be proved that the number of permutations of the sequence $1, \ldots, k$ is $k!$ This will be explained in Lecture 16 (or you can try to prove this on your own). Also, permutations are studied in Lecture 32.

The rearrangement operation takes as input a relation of arity k, rearranges its components, and outputs the resulting relation. A formal definition is this. Let R be a relation of arity k. Let us fix a permutation $i_1, i_2, \ldots, i_k$ of the sequence $1, 2, \ldots, k$. Denote this permutation by α. Then, $\alpha(R)$ is the relation defined as follows:

$$\alpha(R) = \{(a_1, \ldots, a_k) \mid (a_{i_1}, a_{i_2}, \ldots, a_{i_k}) \in R\}.$$

The relation $\alpha(R)$ is called a **rearrangement** of R according to α. For example, consider R_2 in Example 14.1. For the permutation $\alpha = 2, 1$ we have $\alpha(R_2) = \{(1, 0), (2, 0), (3, 0), (4, 0), (3, 2)\}$.

The linkage operation. Suppose we are given two relations R_1 and R_2, of arity s and t, respectively. Let us take a number p such that $0 \le p \le s$ and $p \le t$. We want to define the new relation, $Link_p(R_1, R_2)$.

First, we explain this operation informally. Let us represent R_1 and R_2 as tables. Roughly speaking, $Link_p(R_1, R_2)$ is the table obtained by merging the tables for R_1 and R_2 on the last p columns of R_1 and the first p columns of R_2.

We give an example. Consider the relations R_1 and R_2 in Example 14.1. Let $p = 1$. Then, $Link_1(R_1, R_2)$ is the following ternary relation

$$Link_1(R_1, R_2) = \{(0, 2, 3), (4, 0, 1), (4, 0, 2), (4, 0, 3), (4, 0, 4), (4, 2, 3)\}.$$

We can also link R_1 with U when $p = 1$. In this case,

$$Link_1(R_1, U) = \{(0, 4), (4, 0), (0, 2), (4, 2)\}.$$

Formally, we create a new table, denoted by $Link_p(R_1, R_2)$, that links the tables for R_1 and R_2 as follows. Take a row

$$(a_1, \ldots, a_{s-p}, a_{s-p+1}, \ldots, a_s)$$

in the first table. Note that the length of $(a_{s-p+1}, \ldots, a_s)$ is exactly p. If there is a row $(b_1, \ldots, b_p, \ldots, b_t)$ in the second table such that $(b_1, \ldots, b_p) = (a_{s-p+1}, \ldots, a_s)$, then the tuple $(a_1, \ldots, a_s, b_{p+1}, \ldots, b_t)$ is a row of the new table. Note that the length of this row is $s + t - p$.

Note that when $p = 0$, then $Link_p(R_1, R_2)$ is exactly the Cartesian product of sets R_1 and R_2.

14.4   Exercises

(1) Consider the relational structure in Example 14.1. Write down each of the following relations: $R_1 \cap \overline{R_2}$, $\exists x_1 R_1$, $\exists x_2 R_1$, $c(R_1 \cap \overline{R_2})$, $\exists x_2(R_1 \cup R_2)$, $\exists x_3(c(R_1 \cap R_2))$, and $\exists x_1 \overline{R_2}$.

(2) Let R be a k-place relation. Prove that $\exists x_{k+1} c(R) = R$.

(3) Consider the following relational structure $(D; S, Add, Mult, Div, \leq)$, where

- The domain D is $\{0, 1, 2, 3, 4\}$.
- The successor relation: $S = \{(x, y) \mid y = x + 1 \text{ and } x, y \in D\}$.
- The addition relation: $Add = \{(x, y, z) \mid x+y = z \text{ and } x, y, z \in D\}$.
- The multiplication relation: $Mult = \{(x, y, z) \mid x \cdot y = z \text{ and } x, y, z \in D\}$.
- The order relation: $\leq = \{(x, y) \mid x \leq y \text{ and } x, y \in D\}$.

Do the following:

(a) Write down the tables for S, Add, and $Mult$.

(b) Write down the following relations: $Add \cup Mult$, $\overline{S}$, $\exists x_1 S$, $\exists x_2 S$, $Inst(\leq, 2, 1)$, $Inst(Mult, 1, 1)$, and $\forall x_1 Mult$.

(c) Write down the following relations: $Link_1(S, S)$, $Link_1(Add, S)$, $Link_2(Add, Mult)$.

(d) Write down the following relations: $\exists x_3 Add$, $\exists x_3 Mult$, $\forall x_2 \exists x_3 Add$, $\forall x_2 \exists x_3 Mult$.

Programming exercises

(1) Write a program that given a relation R outputs the cylindrification of R.

(2) Implement the projection and universal projection operations. This should take as input a relation R and an integer i, and output $\exists x_i R$ and $\forall x_i R$.

(3) Given two relations R and Q as well as an integer p, write a method that outputs $Link_p(R, Q)$.

Program correctness through loop invariants

*We are what we repeatedly do. Excellence,
therefore, is not an act but a habit.*

Aristotle.

15.1 Three simple algorithms

A core part of programming is the ability to execute a set of instructions
repeatedly. These repetitions are called *iterations*. Iterations occur in loops,
such as *"for"* and *"while"* loops, which are prevalent in programs. This
lecture aims to develop a mathematical tool, the notion of a loop invariant,
to help us reason about iterations. With our tool at hand, we will learn
how to prove the correctness of algorithms containing iterations.

Proving the correctness of algorithms is obviously important. After all,
we would like to ensure that computer systems used in our daily lives actually work correctly. For instance, we would like air-traffic control systems
to be safe, our banking systems to do correct transactions, nuclear reactor
control systems to not crash. Such systems work according to algorithms
that we design and hence it is of utmost importance that these algorithms
are correct. Below we analyze three simple algorithms that use iterations.
Intuitively, each of these algorithms obviously seems correct. Nevertheless,
their correctness needs to be proved.

Division Algorithm. Recall that the **division theorem** in Lecture 3
states that for any integer $n \geq 0$ and integer $m > 0$, there exists q and r
such that $n = q \cdot m + r$ and $0 \leq r < m$. The Algorithm 15.1 below, on
inputs n and m, produces the required q (quotient) and r (remainder). A
careful reading of the algorithm shows that the $DivisionAlgorithm(n, m)$,

in essence, implements the proof of the division theorem.

Algorithm 15.1 *DivisionAlgorithm*(n, m)

---

(1) Declare variables q and r to be integer variables.

(2) Initialize $q = 0$ and $r = n$.

(3) *While* $r \geq m$ do

    (a) $q = q + 1$.

    (b) $r = r - m$.

(4) Return q, r.

Consider the condition $r \geq m$ before we enter the loop (line 3). The initial value of r is n. If $r \geq m$, then we enter the loop. According to line 3b, the variable r takes the following values consecutively: n, $n - m$, $n - 2m$, $n - 3m$, ... This sequence of positive integers is strictly decreasing. Hence, there must be a stage at which $r < m$. At that stage, the algorithm terminates due to the condition of the *while* loop (line 3). This reasoning shows that for every pair of input integers n and m, where $n \geq 0$ and $m > 0$, the *DivisionAlgorithm*(n, m) terminates. Showing that an algorithm terminates is a necessary and important step in proving its correctness. However, this is not enough. The next question is, how do we prove that the algorithm does in fact output the correct q and r? We discuss this question later in the lecture.

Sorting Algorithm. Let $A[0], A[1], \ldots, A[n-1]$ be an array (or list) of n integers. Each $A[i]$ is called an *item* of the array. The array is **sorted** if $A[0] \leq A[1] \leq \ldots \leq A[n-1]$. We want to solve the following problem, known as **the sorting problem**: Given an array $A[0], A[1], \ldots, A[n-1]$ of n integers, sort the array and output it.

Formally, given an array $A[0], A[1], \ldots, A[n-1]$, we would like to output an array $B[0], B[1], \ldots, B[n-1]$ such that the following properties are satisfied: (1) the array $B[0], B[1], \ldots, B[n-1]$ is sorted, and (2) each item of the array $B[0], \ldots, B[n-1]$ appears in the original array $A[0], \ldots, A[n-1]$ *exactly* as many times as it appears in the sorted array.

The Algorithm 15.2 below, called *SelectionSort*(**A**) algorithm, solves this problem. The algorithm expects array variable **A**$= A[0], \ldots, A[n-1]$ as input.

Informally, the algorithm operates as follows. Run through the array $A[0], A[1], \ldots, A[n-1]$, find the smallest element, and put it into position

---

Algorithm 15.2 *SelectionSort*(**A**) algorithm

---

(1) Initialize $i = 0$.

(2) *While $i \leq n - 1$ do*

    (a) Swap $A[i]$ in the array **A** with the first smallest element in $A[i], \ldots, A[n-1]$.

    (b) Set **A** be the resulting array.

    (c) Increment i.

(3) Output **A**.

---

0 (swapping with whichever item was originally in position 0). Next, run through this newly formed array, but starting from position 1. Again, find the smallest element, but this time, put it into position 1. Repeat this process until the array is exhausted. Once the process terminates, it produces a sorted array. Intuitively, the algorithm is correct but its correctness needs to be proved formally.

Pattern Matching Algorithm. Pattern matching is an important problem in computer science and applications. Demand for finding efficient pattern matching algorithms is dictated by research in information retrieval, query evaluation, program analysis, bio-informatics, etc.

Here is the **pattern matching problem:** Given string $p = p_0 p_1 \ldots p_m$ called the **pattern**, and string $t = t_0 t_1 \ldots t_n$ called the **text**, design an algorithm that either returns all the occurrences of the pattern in the text or a failure message if p does not occur in t.

In order to make this more precise, we say that p **occurs in** t if there exists an i such that $t_i = p_0$, $t_{i+1} = p_1$, $\ldots$, $t_{i+m} = p_m$. In this case, the index i is called an **occurrence index**. There are many algorithms that solve the pattern matching problem. Below we present a straightforward algorithm here in which $t = t_0 \ldots t_n$ and $p = p_0 \ldots p_m$ are input strings.

Algorithm 15.3 *PatternMatching*(p, t) algorithm

---

(1) Initialize $i = 0$.

(2) *While $i + m \leq n$ do*

    (a) if $t_i \ldots t_{i+m} = p_0 \ldots p_m$ then output i.

    (b) Increment i.

(3) Return a failure message if no output has been generated.

Before we discuss the correctness of the three algorithms above, we note that our algorithms are written quite informally without implementation details. Of course, implementing these algorithms requires careful thought. Here we do not discuss issues related to implementation but focus rather on the correctness of the algorithms.

How does one prove that the algorithms above are correct? To answer this question, we analyze some common features of these three algorithms. All three algorithms include an iteration process by using *while* loops. In order to prove the correctness of the algorithms, we must first understand the relationship between the values of the variables before and after iterations. Therefore, we need to analyze the *while* loops carefully.

15.2　The loop invariant theorem

In general, a *while* loop in an algorithm performs an iteration and is of the following form:

$$\text{While } G \text{ do } B$$

The interpretation of this loop is this:

(1) Determine whether G is true.
(2) If G is true then execute B, and then go back to Step 1.
(3) If G is false then exit the loop.

The G part of the loop is called the **guard** and the B part is called the **body** of the loop. The guards in our three examples above are

$$r \geq m, \quad i \leq n - 1, \quad \text{and} \quad i + m \leq n,$$

respectively. A pass through a loop is called an **iteration**. We say that the loop **terminates** if the guard G of the loop becomes false after a finite sequence of iterations. We introduce the following concept that is important in proving the correctness of algorithms.

Definition 15.1. Let S be a statement. We say that the statement S is an **invariant** of the loop

$$\text{While } G \text{ do } B$$

if whenever S is true before an iteration, S remains true after the iteration.

Clearly, statements like $1 = 1$, $n + m = m + n$, and *for each real number x either $x > 0$ or $x = 0$ or $x < 0$* are always true. They are clearly loop invariants. These type of statements are not dependent on the results of iterations. Therefore, they do not tell us anything interesting about the algorithms. Interesting loop invariants are those statements, which depend on the parameters involved in the body of the loop. Finding such invariants are useful because they help us to formally reason about the algorithms and prove their correctness.

We now recast the three algorithms discussed in the previous section, and find their loop invariants. These loop invariants will then be used to prove correctness of the algorithms.

Example 15.1. For the *DivisionAlgorithm(n, m)*, the following statement is a loop invariant:

$$\text{``}n = q \cdot m + r \text{ and } r \geq 0\text{''}.$$

Denote the statement by S. Observe that the statement S involves parameters q and r and inputs n and m. Hence, S is dependent on the execution of the *while* loop in the *DivisionAlgorithm(n, m)*. To see that S is a loop invariant, note that S is true on the first entry into the loop. Indeed, in this case $q = 0$ and $r = n$. Therefore, we have $n = q \cdot m + r$ and $r \geq 0$. Now, assume that S is true before the body of the *while* loop is executed. This means that we have $n = q \cdot m + r$ and $r \geq 0$. After the iteration, the value of q changes to $q' = q + 1$ and the value of r changes to $r' = r - m$. It must be the case that $r' \geq 0$ because otherwise the body would not be executed. Now we do a simple computation: $q' \cdot m + r' = (q+1) \cdot m + (r - m) = q \cdot m + r = n$. Thus, the new values of q and r (which are q' and r' respectively) satisfy the statement S after the iteration of the loop. Therefore, the statement S is a loop invariant.

Example 15.2. For the algorithm *SelectionSort(**A**)*, the following statement is a loop invariant:

"$A[i - 1]$ exceeds no element in the array $A[i], \ldots, A[n - 1]$ and no element in the array $A[0], \ldots, A[i - 2]$ exceeds $A[i - 1]$."

Denote the statement by S. Observe that S has i as a parameter, and thus, again, it depends on the execution of the *while* loop of the algorithm. We now show that S is an invariant of the *SelectionSort(**A**)* algorithm. Clearly, before the first iteration, we do not need to check anything because $A[-1]$ does not exist. So, the statement S is true (vacuously). Before the second iteration, the statement S is true because $A[0]$ is chosen to be the smallest element in the array. Next, assume that S is true before the $(i + 1)$th iteration. During the iteration, the array $A[0], \ldots, A[i - 1]$ is kept intact, and the smallest element among $A[i], \ldots, A[n - 1]$ is put into position i. Therefore, after the iteration, $A[i]$ exceeds no element in the array $A[i + 1], \ldots, A[n - 1]$ and no element in the array $A[0], \ldots, A[i - 1]$ exceeds $A[i]$. Therefore, the statement S stays true after the $(i + 1)$th iteration of the *while* loop.

Example 15.3. For the *PatternMatching*(p, t) algorithm, the following is a loop invariant:

"If $j < i$ is an occurrence index, then j is an output, and otherwise j is not an output."

As in the examples above, S depends on variable i used in the loop. We show that S is an invariant of the loop in the algorithm. Clearly, before the first entry into the *while* loop, S is true (vacuously). Suppose that S is true before the ith iteration, the algorithm has not stopped, and the guard $i + m \leq n$ is true. After the ith iteration, if i is an occurrence index then i is an output, and otherwise i is not an output. Therefore, before iteration $i + 1$ we have the following. If $j < i + 1$ is an occurrence index, then j is an output, and otherwise j is not an output. Thus, S is a loop invariant.

We single out the main property of loop invariants in the following theorem.

Theorem 15.1 (The loop invariant theorem). *Assume that S is an invariant of the loop*

While G do B

Assume also that S is true on the first entry to the loop. Then S stays true after every iteration of the loop. In particular,

(1) If the loop terminates then S is true after the last iteration.
(2) If the loop does not terminate then S always stays true.

Proof.
Our proof is by contradiction. This means that there exists an iteration after which S becomes false. Let k be the first such iteration. Thus,

before each iteration i, where $i = 0, 1, \ldots, k - 1$, the statement S is true. In particular, after iteration $k - 1$, the statement S is true. Moreover, after the $(k-1)^{th}$ iteration, the guard G must be true. Otherwise, the kth iteration would not occur. Since S is an invariant of the loop, the statement S *must* be true after the kth iteration. This is clearly a contradiction as we assumed that S is false after the kth iteration. Thus, S stays true if the loop terminates, and S always stays true if the loop does not terminate. $\square$

15.3   Applications of the loop invariant theorem

The loop invariant theorem plays an important role in proving correctness of algorithms. As an example, we apply the theorem to show the correctness of the three algorithms provided in the first section.

Corollary 15.1. *The algorithm DivisionAlgorithm(n, m) is correct. In other words, for any input integers $n \geq 0$ and $m > 0$, the algorithm outputs q and r such that $n = q \cdot m + r$ and $0 \leq r < m$.*

Proof.   We have already shown, immediately following the description of the algorithm, why the *DivisionAlgorithm* terminates for given inputs n and m. Consider the statement S:

$$\text{``}n = q \cdot m + r \text{ and } r \geq 0\text{''}.$$

In Example 15.1, we have shown that the statement is a loop invariant. Hence the statement S is true after the termination of the algorithm. Note that after the last iteration $r < m$. Thus, the algorithm produces the desired q and r, and is therefore correct. $\square$

Corollary 15.2. *The sorting algorithm SelectionSort$(\mathbf{A})$ is correct.*

Proof.   Clearly, the algorithm terminates after n iterations. Consider the statement S:

"$A[i - 1]$ exceeds no element in the array $A[i], \ldots, A[n - 1]$ and no element in the array $A[0], \ldots, A[i - 2]$ exceeds $A[i - 1]$."

Example 15.2 shows that the statement is a loop invariant. Hence, the statement S is true after the last iteration of the *while* loop. Note that just because the statement is true after the last iteration does not prove that the algorithm produces a sorted array. The fact that the invariant is true for all i is what proves the correctness. Hence, the algorithm produces a sorted array.   □

Corollary 15.3. *The algorithm PatternMatching(p, t) is correct.*

Proof. Consider the following statement S:

"If $j < i$ is an occurrence index then j is an output, and otherwise j is not an output."

Example 15.3 shows that the statement is a loop invariant. Hence, the outputs of the algorithm are *all* occurrence indices of the pattern p in the text t.   □

15.4 Exercises

(1) Consider the *FindFactors* algorithm in Lecture 3. Prove that the algorithm is correct by using a loop invariant. Clearly state what invariant you use.

(2) Consider the *PathExistence* algorithm in Lecture 7. Prove that the algorithm is correct by using a loop invariant. Clearly state what invariant you use.

(3) Prove correctness of Euclidean algorithm using loop invariant theorem. Clearly state what invariant you use.

(4) Explain why the algorithm *PatternMatching*(p, t), for all inputs p and t, terminates.

(5) Discuss the efficiency of the *SelectionSort*$(A[0], \ldots, A([n-1])$ algorithm. For this, assume that it takes one unit of time (1) for swapping items $A[i]$ and $A[j]$, and (2) for comparing the items $A[i]$ and $A[j]$. Your discussion should be based on an analysis of how much time it takes for the algorithm to stop when the size of an input array is n.

Programming exercises

(1) Implement the following algorithms:

   (a) *DivisionAlgorithm*(n, m),
   (b) *SelectionSort*$(A[0], \ldots, A[n-1])$, and
   (c) *PatternMatching*(p, t).

Lecture 16

Induction and recursion

> *To iterate is human, to recurse, divine.*
> L. Peter Deutsch.

16.1 Proofs by induction

In the lectures on trees we saw examples of proofs by induction. In the previous lecture, correctness proofs that used loop invariants were also examples of proofs by induction, even though this was not explicitly stated. In this lecture, our goal is to study proofs by induction in a general framework and learn several inductive proof techniques. These techniques are used to prove properties of natural numbers, algorithms, as well as inductively defined objects such as trees, graphs, and other objects created in programs. These techniques often help in reasoning about the behavior of programs and their correctness.

The first principle of mathematical induction. Suppose that we are given a statement S about natural numbers. Typically, statement S involves a parameter n that acts as a variable for natural numbers. Here is one example of such a statement:

The sum $0 + 1 + 2 + \ldots + n$ of the first n numbers is equal to $n \cdot (n+1)/2$

Another example is this:

Each finite set A with n elements has exactly 2^n subsets.

Both statements involve the parameter n. The first talks about the natural numbers directly. The second talks about a property of sets in terms of the parameter n. These statements are true and their proofs use mathematical induction formulated as follows.

Mathematical Induction Principle I. *Suppose we are given a statement $S(n)$, and that we can prove the following two facts:*

Basis: *The statement $S(0)$ is true.*

Induction Step: *If $S(k)$ is true then $S(k+1)$ is also true.*

Then the statement $S(n)$ is true for all values of n.

In the induction step, in order to carry out a proof by induction, we need to prove the following statement: If $S(k)$ is true then $S(k+1)$ is also true. In proving this statement, we assume that $S(k)$ has already been proved to be true. This is called the **inductive hypothesis**. Then, using the fact that $S(k)$ is true, we prove that $S(k+1)$ is true. We can think of this principle of mathematical induction as a simple version of the loop invariant theorem. Indeed, let us reformulate the following version of the loop invariant theorem (Theorem 15.1).

Theorem 16.1. *Consider the following loop:*

$$\text{while } n \geq 0 \text{ do } \ n = n + 1$$

Now, assume that S is true on the first entry into the loop, and that S is an invariant of this loop. Then S stays true after every iteration of the loop.

Note that in this theorem, the *while* loop never terminates. Let us view the statement S as being dependent on n. The conclusion of the theorem is that S is true for *all* values of n. This is exactly what the first principle of mathematical induction states.

We connect the first principle of mathematical induction and the theorem above as follows:

(1) The basis of the induction principle corresponds to the hypothesis of the theorem that S is true on the first entry to the loop.
(2) The inductive step in the induction principle corresponds to the hypothesis of the theorem that S is a loop invariant of the loop.
(3) The conclusion of the induction principle corresponds to the conclusion of the loop invariant theorem.

Example 16.1. Now we apply the induction principle to the following statement $S(n)$:

The sum $0 + 1 + 2 + \ldots + n$ of the first n numbers is equal to $n \cdot (n+1)/2$.

We want to prove that the statement $S(n)$ is true for all natural numbers n. Here is the proof:

Basis: When $n = 0$, the sum 0 is equal to $0 \cdot (0+1)/2$. Thus, the basis is true.

Inductive Step: The inductive hypothesis is that $S(k)$ is true. In other words, we assume that we have proven that the sum of the first k natural numbers is equal to $k \cdot (k+1)/2$. Then, the sum of the first $k+1$ number is calculated as follows:

$$1 + \ldots + k + (k+1) = (1 + \ldots + k) + (k+1) = k \cdot (k+1)/2 + (k+1) = (k+1) \cdot (k+2)/2.$$

The first equality is just a grouping so that the inductive hypothesis can be used. The second equality is the result of the inductive hypothesis, and the last equality is derived using a simple calculation. Therefore, the statement $S(k+1)$ is true. Thus, by the first principle of mathematical induction, the statement $S(n)$ is true for all natural numbers n.

Example 16.2. Now we apply the induction principle to the next statement $S(n)$:

Each finite set A with n elements has exactly 2^n subsets.

We want to prove that the statement is true for all finite sets. In other words, we want to show that $S(n)$ is true for all natural numbers n. For the proof, recall that $P(A)$ is the power set of A, that is, the set of all subsets of A. Here is an inductive proof of the statement:

Basis: Let A be a set with zero elements. Then $A = \emptyset$, and $P(\emptyset) = \{\emptyset\}$. Thus, $P(A)$ has $2^0 = 1$ elements. That proves the basis.

Inductive Step: The inductive hypothesis is that $S(k)$ is true. In other words, we assume that we have proven that the number of subsets of any set B with k elements is 2^k. Now, assume that a set A has $k+1$ elements. So, say $A = \{a_1, \ldots, a_k, a_{k+1}\}$. Consider the set $B = \{a_1, \ldots, a_k\}$. By the inductive hypothesis, the number of subsets of B is 2^k. Note that all subsets of B are subsets of A. Also, any subset X of A that is not a subset of B must contain a_{k+1}. Since $X \setminus \{a_{k+1}\}$ is a subset of B there must be exactly 2^k subsets of A that are not subsets of B. Since every subset of A is either a subset of B or contains the element a_{k+1}, this implies that the number of subsets of A is

$$2^k + 2^k = 2^{k+1}.$$

Therefore, the statement S is true when $n = k + 1$. By the first principle of mathematical induction, we conclude that S is true for all finite sets.

Example 16.3. Consider the following statement $S(n)$:

The sum $2^0 + 2^1 + 2^2 + \ldots + 2^n$ *is equal to* $2^{n+1} - 1$

We would like to show that the statement is true for all natural numbers $n \geq 0$. When $n = 0$, we indeed have $2^0 = 2^{0+1} - 1$. Hence, the basis is true. Assume that we have proven that $S(k)$ is true. Consider the sum

$$2^0 + 2^1 + \ldots + 2^{k+1}.$$

We group the first $k + 1$ terms of this sum separately, apply the induction hypothesis to the group, and get the following equality:

$$2^0 + 2^1 + \ldots + 2^{k+1} = (2^0 + \ldots + 2^k) + 2^{k+1} = (2^{k+1} - 1) + 2^{k+1} = 2^{k+2} - 1.$$

This proves that the statement $S(k + 1)$ is true. Therefore, the equality

$$2^0 + 2^1 + 2^2 + \ldots + 2^n = 2^{n+1} - 1$$

is true for all natural numbers $n \geq 0$.

The second principle of mathematical induction. In proving a statement $S(n)$ that uses the variable n (for natural numbers), it may well be the case that for some initial values of n, say for the values of n less than a fixed number t, the statement $S(n)$ is false. However, $S(n)$ could be true for all values n greater than t. The number t is called a **threshold**. As a simple example consider the statement $S(n)$:

$$n^2 > n + 16.$$

This statement is false for the values $n = 0, 1, 2, 3, 4$. However, for all $n > 4$, this statement is true. So, $t = 5$ is a threshold. The first principle of the mathematical induction, if applied directly, does not handle this type of situation.

Sometimes, proving $S(n)$ can be even more subtle in the inductive step. The inductive step, in the first principle of mathematical induction, assumes that $S(k)$ is proven to be true. Based on merely this hypothesis alone, the inductive step attempts to prove the statement $S(k + 1)$. However, it may well be the case that in order to prove $S(k + 1)$, the fact $S(k)$ alone may not suffice. We may need to use the fact that *all* statements $S(0)$, $S(1)$, ..., $S(k)$ are proven to be true in order to prove that $S(k + 1)$ is true. We now formalize all this into the second principle of mathematical induction (sometimes called the principle of full induction).

Mathematical Induction Principle II. *Suppose we are given a statement $S(n)$. Let t be a natural number (threshold). Suppose that we can prove the following two facts:*

Basis: *The statement $S(t)$ is true.*

Induction Step: *If $S(t)$, $S(t+1)$, ..., $S(t+k)$ are all true, then $S(t+k+1)$ is also true.*

Then, the statement $S(n)$ is true for all values of $n \geq t$.

Example 16.4. Consider the value $T(i)$ defined as follows:
$$T(i) = \begin{cases} 1 & \text{if } i = 0; \\ 1 + T(0) + T(1) + \ldots + T(i-1) & \text{if } i > 0. \end{cases}$$
For example, $T(1) = 1 + T(0) = 1 + 1 = 2$ and $T(2) = 1 + T(0) + T(1) = 1 + 1 + 2 = 4$. Consider the following statement $S(n)$:

The value of $T(n)$ is equal to 2^n.

We would like to prove that the statement $S(n)$ is true for all natural numbers n. Note that in this case the value of the threshold t is 0. When $n = 0$, we have $T(0) = 1 = 2^0$. Hence, the statement is true in the base case. In the inductive step, assume that all $S(0)$, $S(1)$, ..., $S(k)$ are true. We want to prove the equality $T(k+1) = 2^{k+1}$. Consider the value $T(k+1)$. We have
$$T(k + 1) = 1 + T(0) + T(1) + \ldots + T(k).$$
We use the second principle of mathematical induction. Since all $S(i)$ are true for $i \leq k$, we can replace the equality above by:
$$T(k + 1) = 1 + 2^0 + 2^1 + \ldots + 2^k.$$
Earlier, in Example 16.3 we already proved that $2^{k+1} - 1 = 2^0 + 2^1 + \ldots + 2^k$. Therefore, we have
$$T(k + 1) = 1 + 2^0 + 2^1 + \ldots + 2^k = 1 + (2^{k+1} - 1) = 2^{k+1}.$$

Hence, $S(k+1)$ is true. Therefore, the equality $T(n) = 2^n$ is true for all natural numbers.

16.2   Recursion

Recursion can be used in two different ways. One is to define new objects using what we call recursive definitions. Here, one defines basic objects first, and then, based on precise rules, specifies how to build more complicated objects from ones that are already built. In essence, these types of recursive definitions are the same as inductive definitions. We have already seen several examples of recursive (inductive) definitions: inductive definitions of trees, arithmetic expressions, and labeled trees for arithmetic expressions.

The other way in which recursion is used is for defining functions and algorithms. In this case, a recursively defined function or algorithm F behaves as follows. On a given input, in order to compute the value of F, the function F calls itself on an input of a smaller size. This repeats until F arrives to a base case for which the value of F is defined explicitly. Here is a simple example where the *factorial* function is defined recursively.

(1) If $n \leq 1$, then $factorial(n) = 1$.
(2) If $n > 1$, then $factorial(n) = n \cdot factorial(n-1)$.

To compute $factorial(n)$, the function calls itself to compute $factorial(n-1)$. If $n-1 > 1$, then $factorial(n-1)$ calls itself on $n-2$, and so on. Finally, the value $factorial(1)$ is computed (the base case), which is 1. This value is then returned to $factorial(2)$ which computes 2. The value of $factorial(2)$ is then returned to $factorial(3)$ and so on. Finally, the value $factorial(n-1)$ is returned to $factorial(n)$, which computes $n \cdot (n-1)!$

We give a more interesting example that solves the sorting problem discussed in the previous lecture. We describe the $MergeSort(\mathbf{X})$ algorithm, where $\mathbf{X}$ is an array variable. $MergeSort(\mathbf{X})$ is a solution to the sorting problem which differs from the $SelectionSort(\mathbf{A})$ algorithm.

To describe the $MergeSort(\mathbf{X})$ algorithm, we need to explain a method that we name by $Merge(\mathbf{A}, \mathbf{B})$. The $Merge(\mathbf{A}, \mathbf{B})$ method, given two *sorted* arrays $\mathbf{A}$ and $\mathbf{B}$ of integers, merges these two arrays into one sorted array $\mathbf{C}$. For example, the $Merge(\mathbf{A}, \mathbf{B})$ method, applied to the arrays

$$(2, 2, 5, 6, 6, 9, 10, 11, 11, 11, 14) \text{ and } (1, 1, 3, 4, 4, 8, 9, 12, 12, 14)$$

produces the list $(1, 1, 2, 2, 3, 4, 4, 5, 6, 6, 8, 9, 9, 10, 11, 11, 11, 12, 12, 14, 14)$.

We write the array $\mathbf{X}$ of length t as $(X[0], X[1], \ldots, X[t-1])$. We use the notation $\mathbf{X}^1$ to represent the array obtained, from the nonempty array $\mathbf{X}$, by removing the first element of $\mathbf{X}$, that is $\mathbf{X}^1$ is $(X[1], X[2], \ldots, X[t-1])$. Our $Merge(\mathbf{A}, \mathbf{B})$ algorithm, applied to two arrays $\mathbf{A}$ and $\mathbf{B}$, is the following:

Algorithm 16.1 $Merge(\mathbf{A}, \mathbf{B})$ algorithm

---

(1) If $\mathbf{A}$ is an empty array then output $\mathbf{B}$.
(2) If $\mathbf{B}$ is an empty array then output $\mathbf{A}$.
(3) If $A[0] < B[0]$ then output the array $(A[0], Merge(\mathbf{A}^1, \mathbf{B}))$.
(4) If $B[0] \leq A[0]$ then output the array $(B[0], Merge(\mathbf{A}, \mathbf{B}^1))$.

The algorithm is clearly a recursive algorithm. In the base case, that is when either of the arrays $\mathbf{A}$ or $\mathbf{B}$ is empty, the output is the other array. In the inductive step, the algorithm acts as follows. Firstly, it compares the first items of A and B. Say $A[0] < B[0]$. Secondly, it merges the arrays $(A[1], \ldots, A[n-1])$ with $\mathbf{B}$ thus producing the sorted array $Merge(\mathbf{A}^1, \mathbf{B})$. Finally, the algorithm outputs the array $(A[0], Merge(\mathbf{A}^1, \mathbf{B}))$. The case $B[0] \leq A[0]$ is treated similarly. The reader is left to prove the correctness of this algorithm (Exercise 11).

Having defined the $Merge(\mathbf{A}, \mathbf{B})$ algorithm, we can proceed to the $MergeSort(\mathbf{X})$ algorithm. Let $\mathbf{X}$ be an array $(X[0], X[1], \ldots, X[t-1])$. The **mid point** of $\mathbf{X}$, if t is odd, is $X[k]$, where $2k = t - 1$. If t is even, then the mid point is $X[k]$, where $2 \cdot k = t$. For example, the mid point of $(1, 4, 3, 7, 8, 9)$ is the item 3. The midpoint of $(1, 4, 3, 7, 8, 9, 2)$ is 7.

Now we describe the $MergeSort(\mathbf{X})$ algorithm:

Algorithm 16.2 $MergeSort(\mathbf{X})$ algorithm

---

(1) If $\mathbf{X}$ is the empty array or contains one item only then output $\mathbf{X}$.
(2) If $\mathbf{X}$ contains more than 1 item then do the following:

    (a) Find the midpoint $X[k]$ of $\mathbf{X}$.
    (b) Consider two arrays $\mathbf{A} = (X[0], X[1], \ldots, X[k])$ and $\mathbf{B} = (X[k+1], \ldots, X[t-1])$.
    (c) Set $\mathbf{A} = MergeSort(\mathbf{A})$ and $\mathbf{B} = MergeSort(\mathbf{B})$.

(3) Output $Merge(\mathbf{A}, \mathbf{B})$.

Our goal is to prove, by induction, that the $MergeSort(\mathbf{X})$ algorithm is correct.

Theorem 16.2. *The $MergeSort(\mathbf{X})$ algorithm is correct.*

Proof. Note that $MergeSort(\mathbf{X})$ is clearly a recursive algorithm. When $MergeSort(\mathbf{X})$ is called in line 2c, it is applied to arrays of strictly smaller lengths than that of $\mathbf{X}$.

If the length of $\mathbf{X}$ is 0 or 1, then the first line of the algorithm is applied. In this case, the algorithm outputs $\mathbf{X}$ itself, which is clearly sorted. Assume that the length of $\mathbf{X}$ is $k+1$ and the algorithm correctly sorts *all* the arrays of length at most k. On input $\mathbf{X}$, the algorithm divides $\mathbf{X}$ up into two arrays $\mathbf{A}$ and $\mathbf{B}$, and then applies itself to $\mathbf{A}$ and $\mathbf{B}$. By the induction hypothesis, the arrays $\mathbf{A}' = MergeSort(\mathbf{A})$ and $\mathbf{B}' = MergeSort(\mathbf{B})$, produced in line 2c, are now sorted. Line 3 of the algorithm is then applied to the sorted arrays $\mathbf{A}'$ and $\mathbf{B}'$. Merging $\mathbf{A}'$ and $\mathbf{B}'$ gives the desired sorted array. □

16.3 Exercises

(1) Prove by induction that the number of strings of length n over a binary alphabet is equal to 2^n.
(2) Prove by induction that the number of strings of length at most n, over a binary alphabet, is equal to $2^{n+1} - 1$.
(3) Prove by induction that for all n natural positive numbers, the following equality is true:

$$1^2 + 2^2 + 3^2 + \ldots + n^2 = n \cdot (n+1) \cdot (2n+1)/6.$$

(4) Prove that for all n positive natural numbers, the value $11^n - 4^n$ is divisible by 7.
(5) Prove $n^2 > n + 1$ for all natural numbers $n \geq 2$.
(6) Prove the fundamental theorem of arithmetic (see Lecture 3) using the second principle of mathematical induction.
(7) Prove, by induction, that for every natural number n, there are integers a, b such that $n = 2 \cdot a + 3 \cdot b$.
(8) Prove that the following four statements are equivalent to each other:

    (a) Every nonempty subset of natural numbers contains the smallest number.

(b) Every decreasing sequence $x_1 > x_2 > x_3 > \ldots$ of natural numbers is a finite sequence.

(c) The first principle of mathematical induction.

(d) The second principle of mathematical induction.

(9) Using the second principle of mathematical induction, prove that the $Preorder(T)$ and $Postorder(T)$ algorithms in Lecture 9 print out all the nodes of T without repetition.

(10) This exercise introduces ground terms and asks to define term-trees. Suppose we are given **constant symbols** $c_0, c_1, \ldots, c_m$ and **constructors** $f_0^{n_1}, \ldots, f_s^{n_s}$. The number n_i is called the arity of the constructor $f_i^{n_i}$. The arities of the constructors are non-zero natural numbers. We define **ground terms** as follows.

(a) **Basis:** Every constant symbol is a **ground term**.

(b) **Induction:** : If $t_1, \ldots, t_{n_i}$ are ground terms then the expression $f_i^{n_i}(t_1, \ldots, t_{n_i})$ is also a ground term, where $i \leq s$.

These ground terms can be represented as labeled trees. Do the following:

(a) Give an inductive definition of **ground term trees**.

(b) Prove by induction that the leaves of every term tree are labeled by constants, and the interior nodes by constructors.

(11) Prove by induction that the algorithm $Merge(\mathbf{A}, \mathbf{B})$ is correct.

(12) Discuss the efficiency of the $MergeSort(\mathbf{X})$ algorithm. Compare it with the efficiency of the $SelectionSort((\mathbf{A}))$ algorithm.

Programming exercises

(1) Implement the $Merge(\mathbf{X}, \mathbf{Y})$ algorithm.

(2) Implement the $MergeSort(\mathbf{X})$ algorithm.

Lecture 17

Spanning trees

But I'll tell you what hermits realize. If you go off into a far, far forest and get very quiet, you'll come to understand that you're connected with everything.

Alan Watts.

17.1 Spanning trees

In this lecture, we want to apply our knowledge of trees, graphs and induction from the previous lectures to the analysis of some known algorithms. We pose several interesting problems and provide algorithms that solve these problems.

Our algorithms are motivated by problems that arise in real life. As an informal and not rigorously formulated example, imagine that we have a network of sites that are linked to each other through communication channels. Assume this network is connected, that is, there is always a path between any two sites. The first problem asks to minimize this network while keeping the network connected. The second problem refines the first one as follows. Assume that each communication link has a cost associated with it. We would like to minimize this network to the minimum cost possible while still preserving the connectedness property. Here the minimum cost can, for instance, refer to the sum of costs. These two problems can be transformed and formalized as problems about graphs. In this lecture, we do exactly this and provide algorithms that solve these types of problems. We then prove correctness of these algorithms.

Let $G = (V, E)$ be a graph. Unless otherwise specified, we always assume that our graphs are undirected. Recall from Example 7.2 (in Lecture 7) that a **cycle** in the graph is a path $v_0, \ldots, v_n$ of length greater than 2 such that no vertices in this path are repeated, apart from the start vertex v_0 being equal to the last one v_n. The length of this cycle is n. We now give the following definition of a tree that is somewhat different from the rooted trees studied in Lecture 9.

Definition 17.1. A graph $G = (V, E)$ is a **tree** if G is connected and contains no cycles.

Figure 17.1 presents two examples of trees.

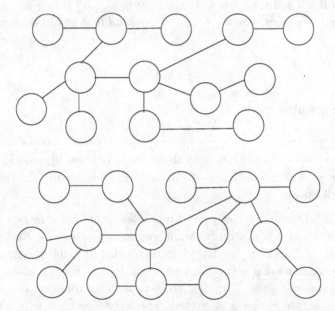

Fig. 17.1: Two examples of trees

Note that the definition of trees does not specify any vertex of the graph to be the root of the tree. It also excludes directions on the edges of the graph. For example, there is no concept of parent and children nodes. However, this definition can *informally* be explained as follows.

Let us select any vertex $v \in V$ in the graph. We pull the vertex v up and let the rest of the vertices dangle down from the selected vertex. Visually, the node v has become the root of the tree and the direction moves down from the root to the leaves. The important part in this visualization is that it is based on the hypothesis that the graph G has no cycles and that G is connected. In this section, by a tree we always mean trees as defined above. The central concept of this section is given in the following definition.

Definition 17.2. Let $G = (V, E)$ be a graph and $T = (V_1, E_1)$ be a tree. We say that T is a **spanning tree** for G if the following two properties are satisfied:

(1) The set of vertices of the graph G and the tree T are the same. In other words, $V = V_1$.
(2) Every edge of the tree T is an edge of the graph G. In other words, $E_1 \subseteq E$.

As an example consider the grid graph represented in Figure 17.2 (a). Figures 17.2 (b), (c), and (d) represent some spanning trees of the grid graph. The thicker black lines represent the edges in the spanning trees.

A natural question arises as to whether every connected graph has a spanning tree. The next theorem gives a positive answer to this question.

Theorem 17.1. *Every connected graph has a spanning tree.*

Proof.
If G is a tree then the theorem is clearly true. Otherwise, the graph G must have a cycle $v_0, v_1, \ldots, v_n, v_0$. Consider the new graph $G_1 = (V_1, E_1)$ with the same vertices and the same set of edges but without the edge $e = \{v_0, v_n\}$. Thus, $V_1 = V$ and $E_1 = E \setminus \{e\}$.

We claim that G_1 is a connected graph. Indeed, let x and y be two vertices in G_1. We want show that there is a path from x to y in the new graph G_1. We know that there is a path

$$q_0, q_1, \ldots, q_m$$

in G such that $q_0 = x$ to $q_m = y$. If this path does not use the edge e (the one we removed) then clearly we still have a path from x to y in G_1. Let us now assume that the path $q_0, q_1, \ldots, q_m$ does use the edge e to move from vertex q_i to vertex q_{i+1}. Since $e = \{v_0, v_n\}$, we may assume that $q_i = v_0$

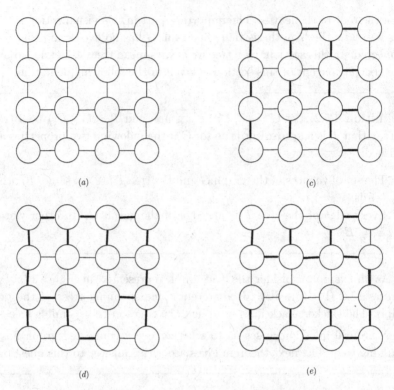

Fig. 17.2: Spanning trees of graph G

and $q_{i+1} = v_n$. We transform the path $q_0, \ldots, q_i, q_{i+1}, \ldots, q_m$ in G to the following path in G_1:

$$q_0, q_1, \ldots, q_i, v_1, \ldots, v_{n-1}, q_{i+1}, \ldots, q_m.$$

This removes one occurrence of the edge e from the original path $q_0, q_1, \ldots, q_m$. We can apply this process repeatedly to all occurrences of e in the original path thus building a path from x to y in the graph G_1. Thus, G_1 is a connected graph.

We replace graph G with G_1 and the set E of edges with E_1 and repeat the same process. The process eventually stops because the set of edges E is a finite set. When the process stops it produces a spanning tree for the graph G because all cycles have now been removed.   □

This theorem suggests Algorithm 17.1 below, called $SpanningTree1(G)$ algorithm, correctly finds a spanning tree for a connected graph $G = (V, E)$.

Algorithm 17.1 *SpanningTree1(G)* algorithm

---

(1) *While* the graph $G = (V, E)$ contains a cycle *do*

   (a) Find an edge e in a cycle.
   (b) Set $V = V$ and $E = E \setminus \{e\}$.
   (c) Set $G = (V, E)$.

(2) Output G.

This algorithm is correct. Correctness is proved using the loop invariant theorem. Indeed, consider the following statement S:

The graph G is connected.

This statement is true on the first entry to the loop. From Theorem 17.1, we see that after every iteration of the loop, S remains true. The loop terminates when the guard *the graph (V, E) contains a cycle* is false. Therefore, when the algorithm terminates it outputs a connected graph with no cycles, that is, a spanning tree for the input graph G.

Though the algorithm above is correct, it is somewhat inefficient because it requires searching for a cycle in the graph after each iteration of the *while* loop.

We now provide a more efficient algorithm, called *SpanningTree2(G, v)* algorithm, for building a spanning tree for a connected graph. The algorithm takes a connected graph G and vertex v as an input. The idea of the algorithm is to build the spanning tree by stages starting with the vertex v. At each stage, the algorithm guarantees that there will be no cycles when we extend the tree built so far.

Algorithm 17.2 *SpanningTree2(G, v)* algorithm

---

(1) Initialize $V_1 = \{v\}$ and $E_1 = \emptyset$.
(2) *While* there is an edge that connects a vertex in V_1 to a vertex not in V_1 *do*

   (a) Find an edge $e = \{u, v'\}$ such that $u \in V_1$ and $v' \notin V_1$.
   (b) Set $V_1 = V_1 \cup \{v'\}$, $E_1 = E_1 \cup \{e\}$, and $G_1 = (V_1, E_1)$.

(3) Output (V_1, E_1).

We prove the correctness of this algorithm in the next theorem.

Theorem 17.2. *If G is a connected graph then the SpanningTree2(G, v) algorithm produces a spanning tree for G.*

Proof.
We first prove that the output of the algorithm is a tree. We again use the loop invariant theorem. Consider the following statement S:

$$\text{The graph } (V_1, E_1) \text{ is a tree that contains node } v.$$

The statement S is true before the first entry into the loop because $V_1 = \{v\}$ and $E_1 = \emptyset$. We now want to show that S is a loop invariant. Suppose that S is true before an entry into the loop. During the iteration, the graph $G_1 = (V_1, E_1)$ is extended as follows:

(1) A new vertex v' is put into V_1.
(2) A new edge $e = \{u, v'\}$ is put into E_1.

After the iteration, our new vertex set V_{new} becomes $V_1 \cup \{v'\}$ and the new edge set E_{new} becomes $E_1 \cup \{e\}$. To prove that S is a loop invariant, we need to show that $G_{new} = (V_{new}, E_{new})$ is a tree that contains v. Clearly, V_{new} contains v. Now, we show that G_{new} contains no cycles. Assume G_{new} does contain a cycle. Then the cycle must use the new vertex v'. Therefore, the degree of v' in G_{new} is at least 2. But, this is impossible because, by the construction of G_{new}, the degree of v' equals 1. Thus, G_{new} is a tree, and hence S is a loop invariant.

Let (V_1, E_1) be the output of the algorithm. To finish the proof we need to show that $V_1 = V$. Clearly, $V_1 \subseteq V$, since the algorithm uses vertices in only V.

In order to arrive to a contradiction, assume that $V_1 \neq V$. This means that there exists a vertex v' in V but not in V_1. Since G is connected there must exist a path from v to v'. Along this path there is an edge $e = \{u, v'\}$ such that $u \in V_1$ and $v' \notin V_1$. This means, the algorithm can not terminate with output V_1 and E_1 because the *while* condition is still active. This contradicts the assumption that (V_1, E_1) is the output of the algorithm. $\square$

17.2 Minimum spanning trees

In this section we extend graphs to graphs with weights. A **weighted graph** consists of a graph $G = (V, E)$ together with weights associated with edges of the graph. If e is an edge in a weighted graph G, then we denote the weight of the edge by $w(e)$. One can interpret the weighted graph in many different ways. For instance, if e is an edge between two vertices x and y, then the weight $w(e)$ might represent the cost of moving between x and y, the cost of communication between x and y, or the time that it takes to move between x and y.

Let G be a weighted graph and T be its spanning tree. **The weight of the tree** T, denoted by $w(T)$, is the sum of the weights of the edges in the tree T. A natural question arises to construct a spanning tree on G such that the weight of the tree is as small as possible. Here is a definition formalizing this concept:

Definition 17.3. We say that a tree T is a **minimum spanning tree** for the weighted graph G if the following two conditions hold:

(1) the tree T is a spanning tree for G, and
(2) the weight $w(T)$ is the least among all the weights of trees spanning G.

As an example consider the weighted graph presented in Figure 17.3 (A). Figures 17.3 (B), (C), and (D) represent spanning trees whose edges are in solid black lines. The weights of these spanning trees are 45, 36, and 23, respectively. So, clearly the spanning tree in Figure 17.3 (D) has the smallest weight among these three spanning trees.

Our goal is to design an algorithm that given a weighted graph G produces a minimum spanning tree for G. There are several algorithms to solve the minimum spanning tree problem. We explain one of them in the next sections.

17.3 Prim's algorithm

Robert C. Prim developed his algorithm for finding minimum spanning trees in 1957. Prim's algorithm is similar to the $SpanningTree2(G, v)$–algorithm where the desired tree is built by stages. Each stage adds an

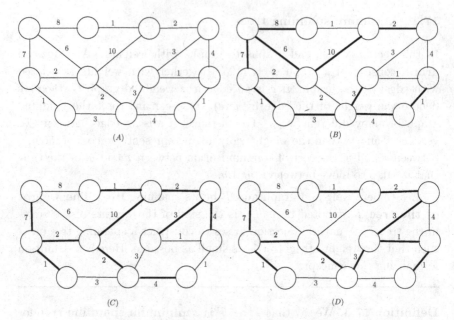

Fig. 17.3: Spanning trees of graph G

edge with the smallest weight to the tree built so far. The idea is that at each stage the algorithm makes a decision that looks optimal. At the end, this sequence of stages produces the desired tree. We name this method the $Prim(G, v)$–algorithm, where G is a weighted graph and v is a vertex of G. Algorithm 17.3 outlines the technique.

Algorithm 17.3 $Prim(G, v)$ algorithm

---

(1) Initialize $V_1 = \{v\}$, $E_1 = \emptyset$, and set $G_1 = (V_1, E_1)$.

(2) *While* there is an edge that connects a vertex in V_1 to a vertex not in V_1 *do*

    (a) Find an edge $e = \{u, v'\}$ with the smallest weight $w(e)$ such that $u \in V_1$ and $v' \notin V_1$.

    (b) Set $V_1 = V_1 \cup \{v'\}$, $E_1 = E_1 \cup \{e\}$, and $G_1 = (V_1, E_1)$.

(3) Output $G_1 = (V_1, E_1)$.

We suggest the reader to run this algorithm on the graph G presented in Figure 17.3 (A). For instance, suppose we start with the vertex placed at the top-left corner of the graph. Namely, we begin with V_1 consisting of this top-left corner vertex. Then the edge e added is the edge with weight 6. Similarly, suppose we reach a stage in the algorithm where V_1 consists of the four vertices that appear in the top part of the graph G. In this case, the edge e added will be the one with weight 3 (adjacent to the node placed on the top-right corner of the graph). Now our goal is to show that the algorithm is correct. This is proved in the next theorem:

Theorem 17.3. *If G is a connected weighted graph then the $Prim(G, v)$-algorithm produces a minimum spanning tree for G.*

Proof.

Let T be the output of the algorithm. We want to show that

(1) T is a spanning tree, and
(2) T is a minimum spanning tree for G.

To prove the first fact we simply repeat the proof of Theorem 17.2. Hence, we can assume that T is a spanning tree. Now we want to show that the weight of T is the least among all the weights of trees spanning G.

First, we make several observations about trees. Take any spanning tree M for G and add a new edge $e = \{u, v\}$ to M. Adding this new edge creates a cycle (see Exercises 1 and 2). Now let us remove any edge e' in this cycle. Then this again results in a spanning tree M'. This is because removing an edge from the cycle does not disconnect the graph. If the deleted edge e' differs from e then this spanning tree M' differs from M.

We now continue our analysis of the $Prim(G, v)$–algorithm. Our goal is to prove that the output tree T is a minimum spanning tree by using the loop invariant theorem. We claim that the following statement S is a loop invariant:

The graph G_1 is a sub-tree of some minimum spanning tree for G.

The statement is true before the first entry into the loop. This is because every weighted connected graph has a minimum spanning tree.

Assume that the statement S is true before an entry into the loop. Hence we assume that there is a minimum spanning tree M for G such that $G_1 = (V_1, E_1)$ is a subtree of M. During the iteration, the graph $G_1 = (V_1, E_1)$ is extended as follows:

(1) A new vertex v' is put into V_1.

(2) A new edge $e = \{u, v'\}$ is put into E_1 such that $w(e)$ is the smallest among all edges such that $u \in V_1$ and $v' \notin V_1$.

After the iteration, we have $V_{new} = V_1 \cup \{v'\}$ and $E_{new} = E_1 \cup \{e\}$. To prove that S is a loop invariant, we need to show that $G_{new} = (V_{new}, E_{new})$ is a sub-tree of a minimum spanning tree.

We would like to construct a new minimum spanning tree M_{new} such that G_{new} is a sub-tree of M_{new}. We construct M_{new} from M. Consider the edge $e = \{u, v'\}$ that we add in this iteration of the loop. If e is an edge in M, then we simply let M_{new} be the same as M, and we are done. If e is not an edge in M, we construct M_{new} as follows.

We add e to M. As we observed earlier, adding e to the tree M produces a cycle. Consider the unique path P from u to v' in M, which does not use e. Note that this path P together with the edge e constitutes the cycle mentioned above. If we trace the path P, we must eventually meet an edge $e_1 = \{x, y\}$ such that $x \in V_1$ and $y \notin V_1$.

Therefore, by the choice of e we have $w(e) \leq w(e_1)$. Otherwise, the algorithm would not have selected e. So, we have added e to M and can now remove e_1 from M. This results in our desired tree M_{new}. The tree M_{new} is still a spanning tree and $w(M_{new}) \leq w(M)$. We conclude that M_{new} is a minimum spanning tree. Clearly M_{new} contains G_{new} as a sub-tree. This proves that S is a loop invariant. Thus, by the loop invariant theorem the output T of the $Prim(G, v)$-algorithm is a minimum spanning tree. □

17.4   Exercises

(1) Prove that if $G = (V, E)$ is a tree then there exists exactly one path from any vertex x in G to any other vertex y in G.

(2) Prove that the following statements for a connected graph $G = (V, E)$ are equivalent:

   (a) The graph G is a tree.
   (b) If an edge is removed from E then the resulting graph is not connected.
   (c) If a new edge is put into E then the resulting graph contains a cycle.

(3) Let T be a tree. Select an edge e in the tree and remove it from T. Show that this results in two disjoint trees T_1 and T_2.

(4) Let T be a minimum spanning tree for G. Select an edge e in the tree and remove it from T. This forms two disjoint trees T_1 and T_2. Consider the graph G_1 defined as follows. The vertices V_1 of G_1 are all vertices of T_1. Edges E_1 of G_1 consists of all edges $e = \{u, v\}$ such that $u, v \in V_1$. Prove that T_1 is a minimum spanning tree for G_1. Similarly, T_2 is a minimum spanning tree for G_2.

(5) Let $G = (V, E)$ be a tree. A **subtree** of G is a tree $G' = (V', E')$ such that $V' \subset V$, and E' is the restriction of E on the set V'. Prove that if $G_1 = (V_1, E_1)$ and $G_2 = (V_2, E_2)$ are subtrees of G and $V_1 \cap V_2 \neq \emptyset$ then $(V_1 \cap V_2, E_1 \cap E_2)$ is also a subtree of G.

(6) The following is Kruskal's algorithm that also finds a minimum spanning tree for connected weighted graphs. On input $G = (V, E)$, where V is a set with n vertices, the $Kruskal(G, v)$–algorithm proceeds as follows:

   (a) Set $T = (V, E_1)$, $E_1 = \emptyset$. (Thus T has n components).

   (b) *While* T has more than one component do

       i. Find an edge e such that $e = \{x, y\}$ has the smallest weight, and x and y belong to two distinct components of T.

       ii. Declare E_1 to be $E_1 \cup \{e\}$.

   (c) Output (V, E_1).

Prove that the algorithm outputs a minimum spanning tree.

Programming exercises

(1) Implement the $SpanningTree2(G, v)$–algorithm.

(2) Implement the $Prim(G, v)$–algorithm.

(3) Implement the $Kruskal(G, v)$–algorithm.

Lecture 18

Shortest paths in directed weighted graphs

Laughter is the shortest distance between two people.

Yakov Smirnoff.

18.1 Shortest paths

Suppose that we are given two vertices s and t in a directed weighted graph G. We call the vertex s the **source** vertex and t the **target** vertex. We would like to find the most optimal path that starts at vertex s and ends at t. A good example of this type of problem is the following. Suppose we have a road map where locations are represented by vertices and the roads between locations are represented by edges. The weights represent the cost required to travel a given road. We would like to find the cheapest way to get from one location to another. This can be transformed into a problem about directed weighted graphs, which is the focus of this lecture.

Let G be a directed weighted graph. Here we always assume that the weights of all edges are non-negative. Let $v_0, v_1, \ldots, v_n$ be a path P. **The weight (or the cost) of this path** P, denoted by $w(P)$, is the sum of the weights of its edges, that is:

$$w(P) = \sum_{i=0}^{n-1} w(e_i),$$

where e_i is the edge from v_i to v_{i+1}. Henceforth, for an edge $e = (x, y)$, we sometimes write $w(x, y)$ instead of $w(e)$.

The shortest path-distance from a vertex u to a vertex v is the minimum weight among the weights of all paths from u to v. We denote this shortest path-distance by $\delta(u, v)$. Thus,

$$\delta(u, v) = \min \{w(P) \mid P \text{ is a path from } u \text{ to } v \text{ in } G\}.$$

177

A **shortest path from u to v** is then any path P whose weight is equal to the shortest path-distance $\delta(u, v)$ from u to v. Note that if negative weights are allowed, then the shortest path-distance from u to v might not exist even if there is a path from u to v. It is an easy exercise for the reader to find such an example.

If there is no path from u to v then we say that the path-distance from u to v is infinite and write this by $\delta(u, v) = \infty$. If there is a path from u to v then the shortest path-distance $\delta(u, v)$ is a real number, and hence we write $\delta(u, v) < \infty$.

Consider the directed weighted graph in Figure 18.1. There are infinitely many paths from p to u. Consider, as an example, four of them: the first path is p, c, u, the second path is p, b, c, u, the third path is p, b, e, f, u, and the fourth path is p, v, c, u. The weights of these paths are 10, 9, 16, and 20 respectively. The path p, b, c, u is the shortest path among these four.

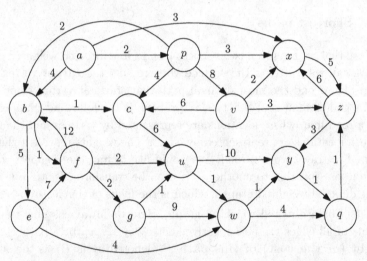

Fig. 18.1: An example of a weighted graph G

18.2   Properties of shortest paths

Let P be a path $v_0, \ldots, v_n$ from u to v in a directed weighted graph G. A **subpath** of P is any continuous segment of P. Thus, all subpaths of P are of the form $v_i, v_{i+1}, \ldots, v_j$, where $i \leq j \leq n$. Note that subpaths of P might have different start and end vertices than those of P.

Theorem 18.1. *A subpath of a shortest path is again a shortest path.*

Proof.
Let P be a shortest path $v_0, \ldots, v_n$ from u to v. We prove the theorem by contradiction assuming that P has a subpath P_1 which is not shortest. So, assume that P_1 is of the form

$$v_i, v_{i+1}, \ldots, v_j.$$

To arrive to a contradiction we use the method called *cut and paste*. We select a path P_2 from v_i to v_j such that $w(P_2) < w(P_1)$. We now cut the path P_1 from P and replace it with the path P_2. The new path P_{new} thus obtained is a path from u to v. Moreover, we have

$$w(u, \ldots, v_i) + w(P_2) + w(v_j, \ldots, v) < w(u, \ldots, v_i) + w(P_1) + w(v_j, \ldots, v).$$

The left side of inequality equals to $w(P_{new})$, and the right side equals $\delta(u, v)$. We have a contradiction with the fact that P is a shortest path from u to v. $\qquad\square$

Theorem 18.2 (Triangle inequality theorem). *For all vertices $x, y, z \in V$ of the directed weighted graph G, the following inequality is true:*

$$\delta(x, z) \leq \delta(x, y) + \delta(y, z).$$

Proof.
For a shortest path P from x to z we obviously have $\delta(x, z) = w(P)$. Let P_1 be a shortest path from x to y and P_2 be a shortest path from y to z. Then $P_1 P_2$ is a path from x to z. In addition $w(P_1) = \delta(x, y)$ and $w(P_2) = \delta(y, z)$. Since $P_1 P_2$ is a path from x to z and P is a shortest path, we have the following inequalities

$$\delta(x, z) = w(P) \leq w(P_1 P_2) = w(P_1) + w(P_2) = \delta(x, y) + \delta(y, z).$$

This is what we wanted to prove. $\qquad\square$

18.3   Formulation of the problem

The shortest path problem is the following. Given a directed weighted graph G and its source vertex s, find the shorest path-distances $\delta(s, v)$ for all vertices $v \in V$. Here are two examples of problems that arise in real life and that can be reduced to the shortest path problem.

A flight agenda example. A travel agency is designing software for making flight plans for its clients. The agency has access to a database of all airports, flights, their costs, and their durations. The agency wants to support both clients that are looking for the cheapest airfares, as well as those that are looking for the fastest way to get from one location to another.

Routing example. Computer networks, such as the Internet, can also be represented as directed weighted graphs. The vertices represent routers, and the weights of edges represent the bandwidth of the connection between routers. The bandwidth is the transmission capacity of a connection. A common problem in computer networks is to find a path, given two vertices, with the highest bandwidth. This problem can be transformed into the shortest path problem with a bit of manipulation on the weights.

18.4   Dijkstra algorithm

In 1959, Edsger Dijkstra designed an efficient algorithm that solved the shortest path problem. In this section, we explain the algorithm. In the next section, we present an example. Before we present the algorithm, we informally describe several of its ingredients. The inputs to the algorithm are a directed weighted graph G and a source vertex s. The algorithm finds the shortest path-distance between s and every vertex in G. The algorithm goes through iterations.

- At each iteration, the algorithm computes an estimate of the path-distances, denoted $d(v)$, from s to all vertices v. If the algorithm has not yet seen a vertex v, then the estimate is set to ∞, which may change in later iterations. Thus, initially $d(s) = 0$, and $d(v) = \infty$ for all other vertices v.

- The algorithm maintains a set F of all vertices whose shortest path-distances from s have already been computed. For instance, $d(s) = 0 = \delta(s, s)$, so s is included in F after the first iteration. Once a vertex v is put into F, its path-distance estimate $d(v)$ is not changed in later iterations and stays equal to $\delta(s, v)$.

- At each iteration, the algorithm puts a vertex $v \in V \setminus F$ into F (thus extending the previous F). When selecting the vertex v, the algorithm picks the one with the smallest path-distance estimate $d(v)$ among the vertices not in F. The intention here is that the value $d(v)$ has now reached the shortest path-distance, and hence v must be added to F.
- Once a vertex v is added to F, the algorithm updates the current path-distance estimates $d(u)$ for all of v's adjacent vertices u. The updating process is called *relaxation*.

Now we present the algorithm, called the $Dijkstra(G, s)$-algorithm. The inputs to the algorithm are a directed weighted graph G and a source vertex s.

Algorithm 18.1 $Dijkstra(G, s)$ algorithm

---

(1) Initialize $d(s) = 0$ and $d(v) = \infty$ for all $v \neq s$. Also, set $F = \emptyset$ and $R = V \setminus F$.
(2) *While* the set R contains a vertex z such that $d(z) < \infty$ do:

    (a) Select $v \in R$ with the smallest path-distance estimate $d(v)$.
    (b) Extend F to $F \cup \{v\}$.
    (c) *Relax* each edge (v, u) outgoing from v as follows. If $d(u) > d(v) + w(v, u)$ then set $d(u) = d(v) + w(v, u)$.
    (d) Remove v from R.

Of course, writing a code to implement the above algorithm requires much thought and time. For instance, one needs to think carefully about how to store and manipulate the elements of F, R and the values $d(v)$ for all the vertices v.

18.5 Example

As an example of how Dijkstra's algorithm works, consider the graph G presented in Figure 18.1. The source vertex is a. Initially $F = \emptyset$, $d(a) = 0$, $R = V$, and $d(v) = \infty$ for all $v \neq a$. The Table 18.1 below shows the steps of the algorithm. The first column of the table represents the number of the iteration. The second column enumerates the elements of the set F. The third column lists the d-values of all vertices in F. The last column presents the d-values of vertices not in the current set F. For the d-values of vertices v that are not in the table, we have $d(v) = \infty$.

#	F	d-values for vertices in F	d-values for vertices not in F
1	a	$d(a) = 0$	$d(p) = 2, d(x) = 3, d(b) = 4$
2	a, p	$d(a) = 0, d(p) = 2$	$d(x) = 3, d(b) = 4, d(c) = 6,$ $d(v) = 10$
3	a, p, x	$d(a) = 0, d(p) = 2, d(x) = 3$	$d(b) = 4, d(c) = 6, d(z) = 8,$ $d(v) = 10$
4	a, p, x, b	$d(a) = 0, d(p) = 2, d(x) = 3,$ $d(b) = 4$	$d(c) = 5, d(f) = 5, d(z) = 8,$ $d(e) = 9, d(v) = 10$
5	a, p, x, b, c	$d(a) = 0, d(p) = 2, d(x) = 3$ $d(b) = 3, d(c) = 5,$	$d(f) = 5, d(z) = 8,$ $d(e) = 9, d(v) = 10, d(u) = 11$
6	$a, p, x, b, c,$ f	$d(a) = 0, d(p) = 2, d(x) = 3$ $d(b) = 3, d(c) = 5, d(f) = 5$	$d(u) = 7, d(g) = 7, d(z) = 8,$ $d(e) = 9, d(v) = 10$
7	$a, p, x, b, c,$ f, u	$d(a) = 0, d(p) = 2, d(x) = 3$ $d(b) = 3, d(c) = 5, d(f) = 5$ $d(u) = 7$	$d(g) = 7, d(z) = 8, d(w) = 8,$ $d(e) = 9, d(v) = 10, d(y) = 17$
8	$a, p, x, b, c,$ f, u, g	$d(a) = 0, d(p) = 2, d(x) = 3$ $d(b) = 3, d(c) = 5, d(f) = 5$ $d(u) = 7, d(g) = 7$	$d(z) = 8, d(w) = 8,$ $d(e) = 9, d(v) = 10, d(y) = 17$
9	$a, p, x, b, c,$ f, u, g, z	$d(a) = 0, d(p) = 2, d(x) = 3$ $d(b) = 3, d(c) = 5, d(f) = 5$ $d(u) = 7, d(g) = 7, d(z) = 8$	$d(w) = 8, d(e) = 9,$ $d(v) = 10, d(y) = 11$
10	$a, p, x, b, c,$ f, u, g, z, w	$d(a) = 0, d(p) = 2, d(x) = 3$ $d(b) = 3, d(c) = 5, d(f) = 5$ $d(u) = 7, d(g) = 7, d(z) = 8,$ $d(w) = 8$	$d(y) = 9, d(e) = 9,$ $d(v) = 10, d(q) = 12$
11	$a, p, x, b, c,$ $f, u, g, z, w,$ y	$d(a) = 0, d(p) = 2, d(x) = 3$ $d(b) = 3, d(c) = 5, d(f) = 5$ $d(u) = 7, d(g) = 7, d(z) = 8,$ $d(w) = 8, d(y) = 9$	$d(e) = 9,$ $d(v) = 10, d(q) = 10$
12	$a, p, x, b, c,$ $f, u, g, z, w,$ y, e	$d(a) = 0, d(p) = 2, d(x) = 3$ $d(b) = 3, d(c) = 5, d(f) = 5$ $d(u) = 7, d(g) = 7, d(z) = 8,$ $d(w) = 8, d(y) = 9, d(e) = 9$	$d(v) = 10, d(q) = 10$
13	$a, p, x, b, c,$ $f, u, g, z, w,$ y, e, v	$d(a) = 0, d(p) = 2, d(x) = 3$ $d(b) = 3, d(c) = 5, d(f) = 5$ $d(u) = 7, d(g) = 7, d(z) = 8,$ $d(w) = 8, d(y) = 9, d(e) = 9,$ $d(v) = 10$	$d(q) = 10$
14	$a, p, x, b, c,$ $f, u, g, z, w,$ y, e, v, q	$d(a) = 0, d(p) = 2, d(x) = 3$ $d(b) = 3, d(c) = 5, d(f) = 5$ $d(u) = 7, d(g) = 7, d(z) = 8,$ $d(w) = 8, d(y) = 9, d(e) = 9,$ $d(v) = 10, d(q) = 10$	

Table 18.1: Iteration process of $Dijkstra$-algorithm

18.6 Correctness of Dijkstra's algorithm

In order to prove that the algorithm is correct, we analyze the algorithm through finding its useful invariants. We start with the following statement S_1:

> *For all $f \in F$ and $r \in R$ we have $d(f) \leq d(r)$.*

Lemma 18.1. *The statement S_1 is a loop invariant.*

Proof.
Before the algorithm enters the loop there is nothing to prove since $F = \emptyset$. After the first iteration, $s \in F$ and $d(s) = 0$. Hence S_1 is true after the first iteration.

Assume that S_1 is true after iteration i. Let F_i and R_i be the values of sets F and R after the iteration. So, we have $d(f) \leq d(r)$ for all $f \in F_i$ and $r \in R_i$. Let v be the vertex added to F_i at iteration $i + 1$. Thus, $F_{i+1} = F_i \cup \{v\}$. Clearly, we have $d(f) \leq d(v)$ for all $f \in F_{i+1}$. Take any vertex $r \in R_{i+1}$. If at iteration $i + 1$ the path-distance estimate $d(r)$ has not been changed then $d(v) \leq d(r)$ by the choice of v. If the path-distance estimate $d(r)$ has been changed, then $d(r) = d(v) + w(v, r)$. This implies that $d(v) \leq d(r)$ for all $r \in R_{i+1}$. Hence $d(f) \leq d(r)$ for all $f \in F_{i+1}$ and $r \in R_{i+1}$. Therefore, S_1 is a loop invariant. $\square$

The proof of this lemma gives us to another useful invariant of the algorithm.

Lemma 18.2. *For each vertex $v \in V$, the path-distance estimate $d(v)$ stays unchanged once v is added to F.* $\square$

By Lemma 18.2, we conclude that the number of iterations of the *while* loop is bounded by the number of vertices of V. Consider now the second statement S_2:

> *For all vertices $v \in V$ we have $\delta(s, v) \leq d(v)$.*

Lemma 18.3. *The statement S_2 is a loop invariant.*

Proof.
The statement S_2 is clearly true before the first iteration. Assume that the statement S_2 is true after iteration i. Let F_i and R_i be the values of sets F and R after the iteration. We want to show that S_2 stays true after iteration

$i + 1$. Consider the vertex v added to F at the iteration. No d-value of $f \in F_{i+1}$, in particular of v, has changed at the iteration by Lemma 18.2. So $\delta(s, f) \leq d(f)$ for all $f \in F_{i+1}$. For all $u \in R_{i+1}$ if the d-value of u has not been changed, then again $\delta(s, u) \leq d(u)$. Otherwise, the edge (v, u) is relaxed at iteration $i + 1$. Therefore, the following inequalities prove the lemma:

$$\delta(s, u) \leq \delta(s, v) + \delta(v, u) \leq \delta(s, v) + w(v, u) \leq d(v) + w(v, u) = d(u).$$

Note that the first inequality follows from the triangle inequality, and the second inequality follows from the fact that $\delta(v, u) \leq w(v, u)$. $\square$

Lemma 18.4. *Let x be a predecessor of y on a shortest path P from s to y. If at some iteration of the algorithm $d(x) = \delta(s, x)$ and the edge (x, y) is relaxed, then $d(y) = \delta(s, y)$.*

Proof.
Since P is a shortest path we have $\delta(s, y) = \delta(s, x) + w(x, y)$. This equality follows from Theorem 18.1. Note that each of the values $d(v)$ is never increased. Therefore, if $d(y) = \delta(s, y)$ before the relaxation is perfomed, then by Lemma 18.3 we are done. So, we assume that $d(y) > \delta(s, y)$ before the relaxation and $d(x) = \delta(s, x)$. Then at the iteration step we have

$$d(y) > \delta(s, y) = \delta(s, x) + w(x, y) = d(x) + w(x, y).$$

The algorithm then sets $d(y) = d(x) + w(x, y) = \delta(s, y)$, which proves the lemma. $\square$

Finally, we are ready to prove the correctness of the Dijkstra's algorithm.

Theorem 18.3. *Dijkstra's algorithm is correct.*

Proof.
We need to show that once the algorithm terminates, the equality $d(v) = \delta(s, v)$ is true for all vertices $v \in V$. From Lemma 18.3, it suffices to prove that each time when a vertex v is added to F, we have $d(v) = \delta(s, v)$.

For vertex s, once s is put into F (at the first iteration), we clearly have $d(s) = 0 = \delta(s, s)$. For the sake of contradiction, suppose the algorithm has reached a stage where it is about to add the first vertex v to F for which $d(v) > \delta(s, v)$. Consider a shortest path P from s to v. Moving from s to v along P, let y be the first vertex not in F. Let x be the predecessor of y along the path P. Note that $x \in F$. Hence, by our assumption $\delta(s, x) = d(x)$.

Since $y \notin F$ when v is added to F we have $d(v) \leq d(y)$. Now consider the iteration when x was added to F. In that iteration we must have:

(1) $d(x) = \delta(s, x)$, and
(2) $d(x) + w(x, y) < d(y)$.

The last inequality holds because the subpath of P from s to y is a shortest path by Theorem 18.1, and $d(x) + w(x, y)$ is the length of that subpath. Hence, the edge (x, y) must be relaxed at that iteration. By Lemma 18.4, we get $\delta(s, y) = d(x) + w(x, y) = d(y)$. Hence, just before v is added to F we must have $d(y) < d(v)$. However, this contradicts the choice of v. $\qquad \square$

18.7 Exercises

(1) Run the $Dijkstra(G, s)$–algorithm on graph G in Figure 18.1 starting with the following values for s: v, p, b and q.
(2) $Dijkstra(G, s)$–algorithm is applied to graphs with non-negative weights. Why do we need this assumption?
(3) $Dijkstra(G, s)$–algorithm does not produce the shortest path from s to v but rather outputs the shortest distance $\delta(s, v)$ for all v in G. Change the algorithm to output $d(v)$ together with a shortest path from s to v.
(4) In the proof of Theorem 18.3 we state the following:
  (1) $d(x) = \delta(s, x)$, and
  (2) $d(x) + w(x, y) < d(y)$.

  Explain in more details why these statements are true.

Programming exercises

(1) Implement $Dijkstra(G, s)$–algorithm.
(2) Extend your implementation of the algorithm so that in addition to $\delta(s, v)$, it also outputs a shortest path from s to v, for all v in G.

Games played on finite graphs

It's not so important who starts the game but who finishes it.

John Wooden.

19.1 Bipartite graphs

Imagine a scenario where the user or a program competes against a program. For example, two robots compete to reach a certain region on a map while preventing the other from doing so. Another example is when the user plays against a chess program. We can think of the competitors as players, where one player attempts to beat the other by reaching a certain state. The aim of this lecture is to formalize these types of games using graphs and to design algorithms that determine the winners of such games.

We need to define a special class of directed graphs called **bipartite graphs**. These graphs are well-suited for representing two player games.

Definition 19.1. A **bipartite graph** is a directed graph $G = (V, E)$ with two sets of vertices V_0 and V_1 such that the following properties are true:

(1) $V_0 \cap V_1 = \emptyset$ and $V_0 \cup V_1 = V$.
(2) The set E of edges is a subset of $V_0 \times V_1 \cup V_1 \times V_0$.
(3) For every vertex $v_0 \in V_0$ there exists a vertex $v_1 \in V_1$ such that $(v_0, v_1) \in E$.
(4) For every vertex $v_1 \in V_1$ there exists a vertex $v_0 \in V_0$ such that $(v_1, v_0) \in E$.

Part (1) stipulates that every vertex is either in V_0 or V_1 but never both.

Part (2) of the definition tells us that no edge connects two vertices from the same set V_i. Parts (3) and (4) say that there is an outgoing edge from any vertex of the bipartite graph. Sometimes, we write the bipartite graph G as $G = (V_0 \cup V_1, E)$ to single out the sets V_0 and V_1.

Example 19.1. Figure 19.1 illustrates an example of a bipartite graph, where $V_0 = \{a, b, c, d, e, f, i, j, k, m\}$ and $V_1 = \{0, 1, 2, 3, 4, 5, 6, 7\}$. We leave it up to the reader to verify that the graph in Figure 19.1 is indeed a bipartite graph, with V_0 and V_1 specified as above.

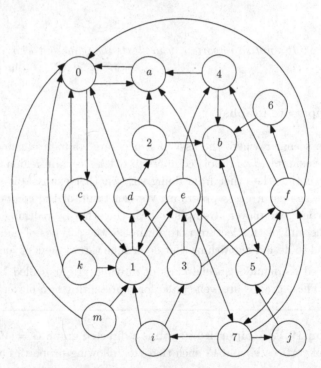

Fig. 19.1: An example of a reachability game

19.2   Reachability games

The games we want to define occur on bipartite graphs $G = (V_0 \cup V_1, E)$. In these games there are two players:

Player 0   and   *Player* 1.

The vertices in V_0 are called the **positions** of *Player* 0. The vertices in V_1 are called the **positions** of *Player* 1. A play between these two players is described as follows. The play starts at any vertex v_0. Say the vertex is in V_0. In this case, *Player* 0 selects an edge $e = (v_0, v_1)$, moves along the edge, and passes control to the opponent. Then, *Player* 1 selects an edge $e = (v_1, v_2)$, moves along the edge and passes control to the opponent. This continues turn by turn. The last two parts of the definition for bipartite graphs imply the following. At any given stage of the play, the players are able to make moves and continue the play. We formally define the concept of a play in the following definition.

Definition 19.2. A **play** in a bipartite graph $G = (V_0 \cup V_1, E)$ is a (finite or infinite) sequence of vertices

$$v_0, v_1, v_2, \ldots,$$

such that (v_0, v_1), (v_1, v_2), (v_2, v_3), ... are all edges of the graph.

Examples of infinite plays in the bipartite graph in Figure 19.1 are

$$0, c, 1, c, 0, c, 1, c, 0, c, 1, c, \ldots \quad \text{and} \quad f, 4, b, 5, f, 4, b, 5, f, 4, b, 5, \ldots.$$

Initial parts of these plays are examples of finite plays.

Now we introduce the notion of a win for a player. To do this, we select a subset of vertices $T \subseteq V$ in the bipartite graph G. The set T is viewed as the region that *Player* 0 wants to reach. So, the goal of *Player* 0 is to reach the set T, whereas the opponent's goal is to keep *Player* 0 out of T. The set T is called the **target set** for *Player* 0. Keeping this in mind, we give the following definition.

Definition 19.3. A **reachability game** Γ consists of:

(1) The bipartite graph $G = (V_0 \cup V_1, E)$.
(2) The target set T of vertices $T \subseteq V_0 \cup V_1$.
(3) Two players: Player 0 and Player 1.

We say that Player 0 **wins a play**

$$v_0, v_1, v_2, \ldots$$

if there exists an i such that $v_i \in T$. Otherwise, Player 1 wins the play.

To give a better intuition for the definitions that we give in the next section, we analyze the reachability game played on the graph in Figure 19.1, where:

(1) The set of *Player 0*'s positions is $V_0 = \{a, b, c, d, e, f, i, j, k, m\}$,
(2) The set of *Player 1*'s positions is $V_1 = \{0, 1, 2, 3, 4, 5, 6, 7\}$, and
(3) The target set T for *Player* 0 is $T = \{a, b\}$.

As an example, consider the two plays that we discussed earlier. The first play was $0, c, 1, c, 0, c, 1, c, 0, c, 1, c, \ldots$. In this case, Player 1 wins because the play does not visit the target set T (namely, neither a nor b appears in the play). The second play was $f, 4, b, 5, f, 4, b, 5, f, 4, b, 5, \ldots$. In this case, Player 0 wins the play.

Suppose now that the game starts at some vertex v. We want to know whether *Player* 0 can reach the target set T independent of Player 1's moves, given that the play starts at v. We reason as follows.

Consider the case where the game starts with $v = a$ or $v = b$. Both are winning vertices for *Player* 0 because every play starting from these vertices, by definition, is won by *Player* 0. In other words, all the plays starting from any of these two vertices are won by *Player* 0 no matter what the opponent does.

Next, consider the vertices 2, 4, and 6 (recall that these are Player 1's vertices). If the game starts from any of these vertices, then no matter where *Player* 1 moves, all of the moves end in the target set T. Therefore, these vertices are also winning for Player 0.

Consider the vertices d, e, and f (these are Player 0's vertices). If the game starts at vertex d then *Player* 0 can guarantee a win by moving to vertex 2 since we already know that 2 is a winning vertex. If the game starts at e then *Player* 0 can guarantee a win by moving to vertex 4 since we already know that 4 is a winning vertex. Similarly, if the game starts at f then *Player* 0 can guarantee a win by moving to vertex 6 since we already know that 6 is a winning vertex. Thus, all the vertices d, e, and f are winning for *Player* 0.

Consider vertices 3 and 5. If the game starts at 3 then Player 1 is forced to move into either d, e or f. We already know that d, e, and f are winning vertices for *Player* 0. Therefore, 3 is also winning for *Player* 0. Similarly, if the game starts at 5 then no matter what move *Player* 1 makes, the player moves to either b, e or f. We already know that b, e, and f are winning vertices for *Player* 0. Therefore, 5 is also winning for *Player* 0.

Let us analyze the vertex 0. If *Player* 1 moves to a then the player looses. Thus, *Player* 1 moves to c. At vertex c, *Player* 0 has two choices: one is to move back to 0 and the other is to move to 1. If *Player* 0 moves back to 0 then *Player* 1 repeats the previous move. Suppose that *Player* 0 moves to 1. In this case, *Player* 1 moves back to c. Thus, if Player 1 keeps moving back to c from both 0 and 1, then no matter what *Player* 0 does, no play reaches T. Therefore, if the game starts from vertex 0, then *Player* 1 can keep out of T forever.

The reader can now check that for the reachability game played on the graph presented in Figure 19.1 with $T = \{a, b\}$, the following vertices are the winning vertices for *Player* 0:

$$a, \ b, \ 6, \ 2, \ 4, \ d, \ e, \ f, \ 3, \ 5, \ i, \ j, \ \text{and } 7.$$

The following vertices are winning for *Player* 1: $0, c, 1, k, m$. Indeed, if the game starts from any of these vertices, then *Player* 1 can always guarantee that the target set T is never reached by *Player* 0. Note that no vertex is winning for both of the players.

19.3   Solving reachability games

Our goal is the following. Given a reachability game, we would like to design a method that finds all the positions from which Player 0 can win the game not depending on the moves made by the opponent. We explain this with the following more formal definition.

Definition 19.4. Let Γ be a game given by the bipartite graph $G = (V_0 \cup V_1, E)$ and a target set $T \subseteq V_0 \cup V_1$. Assume that the game starts from a vertex v.

(1) The vertex v is **winning** for Player 0 if the player can reach the set T no matter what moves the opponent makes.
(2) Similarly, the vertex v is **winning** for Player 1 if the player can keep the plays out of T forever no matter what moves the opponent makes.

Our aim is to design an algorithm that, given a reachability game Γ, determines:

(1) all the vertices that are winning for Player 0, and

(2) all the vertices that are winning for Player 1.

In addition, our discussions below will show that *every* vertex is winning either for *Player* 0 or *Player* 1.

Let Γ be a reachability game given by $G = (V_0 \cup V_1, E)$ and the target set $T \subseteq V_0 \cup V_1$. We define, by induction, the sequence $X_0, X_1, X_2, \ldots$ of subsets in $V = V_0 \cup V_1$ and the **ranks** of vertices as follows:

Basis. Set $X_0 = T$. For all $v \in X_0$, declare $rank(v) = 0$.

Inductive Step. Our induction hypothesis assumes that we have the following:

(1) We have built sets $X_0, X_1, \ldots, X_n$.
(2) For every vertex v, if $v \in X_i$ then $rank(v) = i$. If v does not belong to any of the sets $X_0, X_1, \ldots, X_n$, then $rank(v)$ has not yet been defined.

Now we want to build the set X_{n+1}. For every vertex v whose rank has not yet been defined we proceed as follows. There are two cases.

Case 1: $v \in V_1$. In this case, we set $rank(v) = n + 1$ if for *every* edge (v, u) outgoing from v it must be the case that $u \in X_0 \cup \ldots \cup X_n$. If $rank(v) = n + 1$, we put v into X_{n+1}.

Case 2: $v \in V_0$. In this case, we set $rank(v) = n + 1$ if *there exists* an edge (v, u) outgoing from v such that $u \in X_0 \cup \ldots \cup X_n$. If $rank(v) = n+1$, we put v into X_{n+1}.

If $X_{n+1} = \emptyset$, then we stop the process of building the sets $X_0, \ldots, X_n$. Note that if $X_{n+1} = \emptyset$ then no vertex v exists such that $rank(v) \geq n + 1$.

We now explain and analyze this inductive process in the following example.

Example 19.2. Consider the reachability game where the graph G is the one in Figure 19.1 and the target set $T = \{a, b, 6\}$. The above process starts with $X_0 = \{a, b, 6\}$, and produces the following sequence of sets: $X_1 = \{4, 2, f\}$, $X_2 = \{d, e\}$, $X_3 = \{3, 5\}$, $X_4 = \{i, j\}$, $X_5 = \{7\}$, and finally $X_6 = \emptyset$. All the vertices in $X_0 \cup X_1 \cup X_2 \cup X_3 \cup X_5$ are assigned ranks. The vertices 0, c, 1, k, and m have no ranks. In addition, one can check the following. If $rank(v) = i$, where $i = 1, 2, 3, 4, 5$, then:

(1) If $v \in V_1$ then *all* moves from v made by *Player* 1 goes to a vertex of a strictly smaller rank.

(2) If $v \in V_0$ then *there is* a move from v made by *Player* 0 that goes to a vertex of rank $i - 1$.

Therefore, from all of the ranked vertices, *Player* 0 can reach the target set T. Hence, ranked vertices are winning for *Player* 0. Moreover, unranked vertices v have the following properties:

(1) If $v \in V_1$ then *there is* a move from v made by *Player* 1 that goes to an unranked vertex.
(2) If $v \in V_0$ then *all* moves from v made by *Player* 0 go to unranked vertices.

Note that unranked vertices are not in T. Therefore, from all of the unranked vertices, *Player* 1 can keep plays out of T forever. Hence, unranked vertices are winning for *Player* 1. Thus, we have completed our analysis of the reachability game Γ played on the graph G presented in Figure 19.1 with $T = \{a, b, 6\}$.

Now our goal is to prove, more generally, that ranked vertices are winning for *Player* 0, and unranked vertices are winning for *Player* 1 in any reachability game Γ.

Theorem 19.1. *Let Γ be a reachability game given by the bipartite graph $G = (V_0 \cup V_1, E)$ and the target set $T \subseteq V$ (where $V = V_0 \cup V_1$). Then:*

(1) all the ranked vertices are winning for Player 0, and
(2) all the unranked vertices are winning for Player 1.

In particular, every vertex must be winning for either Player 0 *or* Player 1 *but not both.*

Proof.
We prove the first part of the theorem by induction on the ranks of the vertices. Let v be a ranked vertex.

Base case. Assume that $rank(v) = 0$. Then, by the definition of X_0, we have $v \in X_0$. Since $X_0 = T$, any play from v is won by *Player* 0.

Now our inductive assumption is this:

All the vertices q such that $rank(q) \leq k$ are winning for *Player* 0.

Now, let the rank of v be $k + 1$. Using the inductive hypothesis, we want to show that v is a winning vertex for *Player* 0. For the vertex v there are two cases.

Case 1: $v \in V_1$. In this case, for *every* edge (v, u) it must be the case that $rank(u) \leq k$. Thus, all moves of *Player* 1 from v go to vertices of strictly smaller ranks. All the vertices of smaller ranks are winning for Player 0, by the inductive hypothesis. Therefore, no matter where *Player* 1 moves from v, the move goes to a vertex of a smaller rank that is a winning vertex for *Player* 0. Hence, v is a winning vertex for *Player* 0.

Case 2: $v \in V_0$. In this case, there must exist an edge (v, u) such that $rank(u) = k$ since $rank(v) = k + 1$. Thus, Player 0 does the following. From vertex v, the player moves to u. Since $rank(u) = k$, we can apply the inductive hypothesis by which Player 0 can guarantee a win from u. Therefore, v is a winning vertex for Player 0.

Thus, we have proved that every ranked vertex is a winning vertex for *Player* 0.

Now we prove the second part of the theorem. Note that all unranked vertices are *not* in T. Therefore, it suffices to show that Player 1 can stay in unranked vertices once the game starts in an unranked vertex. Thus, let v be an unranked vertex. For the vertex v there are two cases.

Case 1: $v \in V_1$. In this case, *there exists* an edge (v, u) such that u is unranked. Otherwise, v had to be ranked. Therefore, *Player* 1 simply moves to any of these unranked vertices.

Case 2: $v \in V_0$. In this case, for *every* edge (v, u) it must be the case that u is unranked. Otherwise, v had to be ranked. Therefore, no matter where *Player* 0 moves, the move has to go to an unranked vertex.

Thus, if the game starts in an unranked vertex, then *Player* 1 can keep plays out of T by just staying in unranked vertices. Hence, every unranked vertex is winning for *Player* 1. $\qquad \Box$

19.4 Exercises

(1) Consider the bipartite graph in Figure 19.1. On this graph, consider the reachability games for the following values of T: $\{a, b\}$, $\{7\}$, $\{e, 3, 5\}$,

$\{a, 1\}$, and $\{0, a, b\}$. For each of these reachability games, find the winning vertices for *Player* 0 by computing their ranks.

(2) Let Γ be a reachability game on a bipartite graph $G = (V_0 \cup V_1, E)$ with set $T \subseteq V$. Let $v_0, \ldots, v_k$ be a play. We say that this play is produced by k moves made by the players. Prove that $rank(v) = n$ if and only if *Player* 0 can reach set T in at most n moves made by the players no matter what moves the opponent makes.

(3) Write down an algorithm that takes as input a reachability game and outputs the winning vertices for *Player* 0 and winning vertices for *Player* 1.

(4) Let G be a reachability game. Fix a positive integer n. We say that a vertex v is n-**winning** if Player 0 can guarantee that the target set T is reached at least n times no matter what the opponent does. Design an algorithm that finds all n-winning vertices. Prove the correctness of your algorithm.

(5) A **reach and avoid game** Γ consists of:
   (a) The bipartite graph $G = (V_0 \cup V_1, E)$.
   (b) The target set T of vertices $T \subseteq V_0 \cup V_1$.
   (c) The avoid set $A \subseteq V_0 \cup V_1$ such that $T \cap A = \emptyset$.
   (d) Two players: Player 0 and Player 1.

   We say that Player 0 **wins a play**

   $$v_0, v_1, v_2, \ldots$$

   if there exists an i such that $v_i \in T$ and all $v_0, v_1, \ldots, v_{i-1}$ are not in A. Otherwise, Player 1 wins the play. In other words, Player 0 wins the play if the play visits T before A is visited. Design a method that given a reach and avoid game, finds all positions of the game from which Player 0 can win the game.

Programming exercises

(1) Implement the algorithm from Exercise 3.
(2) Implement the algorithm from Exercise 4.
(3) Implement the algorithm from Exercise 5.

Lecture 20

Functions

*The important thing in science is not so much to obtain new
facts as to discover new ways of thinking about them.*

William Bragg.

20.1 Definition of function

In this lecture, we study functions. This term is quite familiar to us because,
for instance, we have already seen functions that add, multiply and subtract
integers. Informally, a function is a transformation that can be applied to
objects of a certain type, in order to produce objects of the same or different
type. We now give a more formal definition.

Definition 20.1. A **partial function** is given by the following three objects:

(1) A set A called the **input** set of the function.
(2) A set B called the **output** set of the function.
(3) A rule f that transforms some elements of A to some elements of B
    such that no element a from A is transformed to more than one element
    of B.

A shorthand for the partial function is $f : A \to B$ or simply f itself. If f
transforms a to b then we write this as $f(a) = b$.

Given a partial function $f : A \to B$, we say that f maps the set A into
the set B. If $f(a) = b$ then we say that b is the **output** of the partial
function on **input** a. We collect all the outputs of the partial function f.

This set is called the **range** of the partial function $f : A \to B$, and is denoted by $Range(f)$. Formally:

$$Range(f) = \{b \mid \text{there is an input } a \in A \text{ such that } f(a) = b\}.$$

The definition of a partial function f does *not* require that f gives an output for *every* input a. This is the reason why these are called "partial" functions. If the partial function f does not give an output for an input a, then we say that f *is undefined on a* or f *diverges on a*.

Example 20.1. Consider the partial function $f : \mathbb{N} \to \mathbb{N}$ defined as follows. If $n > 0$ then $f(n) = n - 1$. This partial function is clearly defined on all positive integers. It is undefined on 0, since we did not specify the value of f on input 0. The range of the partial function f is $\mathbb{N}$ itself. This is because for every $n \in \mathbb{N}$ it is the case that there exists some m such that $f(m) = n$, clearly, $m = n + 1$.

For a partial function $f : A \to B$, we can also collect all the inputs $a \in A$ for which $f(a)$ is defined. This collection (of such inputs) is called the **domain** of the partial function. Formally:

$$Domain(f) = \{a \in A \mid \text{the output } f(a) \text{ is defined}\}.$$

Thus, $Domain(f)$ does not contain elements of A on which f is undefined.

Example 20.2. Consider the partial function $f : \mathbb{N} \to \mathbb{N}$ defined in Example 20.1 above. The domain of the partial function f is then $Domain(f) = \{n \in \mathbb{N} \mid n > 0\}$. Clearly, the domain does not contain 0.

Definition 20.2. If a partial function $f : A \to B$ gives output to *every* input then we say that f is a **total function** or simply a **function**. Thus, the term "function" without the adjective partial will always mean "total function".

If a partial function $f : A \to B$ is total then its domain and input set A are equal. Every function is a partial function. We denote partial functions and functions with letters f, g, h, and so on. Here are several examples:

Example 20.3.

(1) The partial function $f : \mathbb{Z} \to \mathbb{Z}$, where $f(n) = -n$. The domain and the range of this function are both $\mathbb{Z}$. In our notations, $Domain(f) = \mathbb{Z}$ and $Range(f) = \mathbb{Z}$.

(2) The partial function $S : \mathbb{N} \to \mathbb{N}$, where $S(n) = n + 1$. This function is often called the **successor function**. The domain of this function is $\mathbb{N}$ and the range is the set of all positive integers. In our notations, $Domain(S) = \mathbb{N}$ and $Range(S) = \{n \in \mathbb{N} \mid n > 0\}$.

(3) The partial function $sign : \mathbb{Z} \to \mathbb{Z}$, where $sign(n) = -1$ if n is negative, $sign(n) = 1$ if n is positive, and $sign(n) = 0$ if $n = 0$. The domain of this function is $\mathbb{Z}$, and the range is $\{-1, 0, 1\}$. So,

$$Domain(f) = \mathbb{Z} \text{ and } Range(f) = \{-1, 0, 1\}.$$

(4) The partial function $f : \mathbb{Z} \to \{a, b, x\}$, where $f(n) = a$ if n is even; $f(n) = b$ if n is odd and positive. This function is undefined on all numbers that are odd and negative. Thus,

$$Domain(f) = \mathbb{Z} \backslash \{n \mid n \text{ is an odd negative integer}\}, \; Range(f) = \{a, b\}.$$

The first three examples of partial functions above are (total) functions. Also, one common property of all the partial functions in this example is that either the input set or the output set of the partial function is infinite. We single out these in the next definition:

Definition 20.3. An **infinite partial function** is one whose input or output sets are infinite. Thus, a **finite partial function** is one whose input and output sets are both finite.

Tables can be used to describe finite partial functions. The table lists all inputs and corresponding outputs. Here is an example.

Example 20.4. Consider the partial function

$$f : \{a, b, c, d, e\} \to \{1, 2, 3, 4, 5\}$$

presented in the table below.

a	b	c	d	e
2	3	5	2	

The first row of the table lists all the inputs and the second row lists corresponding outputs. Thus, $f(a) = 2$ and $f(c) = 5$ for example, and $f(e)$ is undefined. The domain of this partial function is $\{a, b, c, d\}$. The range of this partial function is $\{2, 3, 5\}$.

The set of inputs for a partial function may consist of tuples, sequences, or different types of objects. For example, the set of inputs for the addition function on integers is the collection of all 2-tuples (n, m) of integers. Here is another example.

Example 20.5. Consider the partial function

$$f : \{1, 2, 3, 4\} \times \{a, b, c\} \rightarrow \{0, 1, 2, 3\}.$$

The partial function f is represented in the table below.

	a	b	c
1	0	2	2
2	2	1	
3	0		1
4	0	2	1

The input set for f is $\{1, 2, 3, 4\} \times \{a, b, c\}$ and the output set is $\{0, 1, 2, 3\}$. For example, we have $f(1, b) = 2$, $f(3, c) = 1$, and $f(3, b)$ is undefined. It is clear that for this partial function f we have the following: $Domain(f) = \{1, 2, 3, 4\} \times \{a, b, c\} \setminus \{(3, b), (2, c)\}$ and $Range(f) = \{0, 1, 2\}$.

In the last two examples, the input and output sets of the partial functions are finite sets. Therefore, these functions are finite.

20.2 Equality of functions

The concept of equality between partial functions is an important one. Given two partial functions f and g, it does not suffice to check if f and g give the same output to a particular input in order to determine whether f and g are equal. The equality should be checked for *all* inputs.

Definition 20.4. We say that a partial function $f : A \rightarrow B$ **equals to** another partial function $g : A' \rightarrow B'$ if each of the following is true:

(1) $A = A'$ and $B = B'$.
(2) For every $a \in A$, the value $f(a)$ is defined if and only if the value $g(a)$ is defined.
(3) For every $a \in A$, if $f(a)$ is defined then $f(a) = g(a)$.

If f and g are equal then we denote this by $f = g$.

If partial functions $f : A \to B$ and $g : A' \to B'$ are finite and given explicitly, in other words if the elements of A, A', B and B' are listed and f and g are presented by tables, then we can determine whether or not f and g are equal. This can be done by applying the following algorithm step by step:

Algorithm 20.1 *EqualityCheck(f, g)* algorithm

---

(1) Determine if $A = A'$ and $B = B'$. If any of these equalities is false then output $f \neq g$.

(2) Determine if there is an $a \in A$ such that $f(a)$ is defined and $g(a)$ is not. If so, then output $f \neq g$.

(3) Determine if there is an $a \in A$ such that $g(a)$ is defined and $f(a)$ is not. If so, then output $f \neq g$.

(4) Determine if there is an $a \in Domain(f)$ such that $f(a) \neq g(a)$. If there is such an a then output $f \neq g$. Otherwise, output $f = g$.

An interesting observation is concerned with determining the equality of two partial functions $f : A \to B$ and $g : A' \to B'$ when their input sets are infinite. A simple case is when both the input and the output sets for f and g are $\mathbb{N}$ and when f and g are both total functions. How difficult is it to determine if $f = g$? We briefly analyze this question below.

Say we are given two functions $f : \mathbb{N} \to \mathbb{N}$ and $g : \mathbb{N} \to \mathbb{N}$. Note that these functions are total (because we did not say "partial"). Assume we would like to determine whether these two functions are equal. To do that let us consider the following while loop:

While $n \geq 0$ *do*
if $f(n) = g(n)$ then $n = n + 1$. Otherwise, output $f \neq g$ and stop.

The equality of these two functions f and g is then equivalent to saying that the *while* loop never terminates. Indeed, if $f \neq g$ then there must exist an n such that $f(n) \neq g(n)$. Thus, if we choose n to be the minimal number n such that $f(n) \neq g(n)$ then the loop must stop at the $(n+1)^{th}$ iteration. On the other hand, if $f = g$ then $f(0) = g(0)$, $f(1) = g(1)$, $f(2) = g(2)$, and so on. Hence the loop never terminates. This observation suggests that proving the equality of two functions whose domain is $\mathbb{N}$ (or any other infinite set) may involve the use of the induction principles and a good understanding of the functions.

20.3　Composition of functions

In this section we consider total functions only. The composition operation is a method that takes as input two functions and outputs another function. A formal definition of this operation is the following. Suppose we are given two functions

$$f : A \to B \quad \text{and} \quad g : B \to C.$$

The **composition** of these two functions is a function, denoted by $g \circ f$, defined as follows:

(1) The set of inputs of the function $g \circ f$ is the set A.
(2) The set of outputs of the function $g \circ f$ is the set C.
(3) The function $g \circ f$ transforms every element $a \in A$ into $g(f(a))$. Formally, for all $a \in A$:

$$g \circ f(a) = g(f(a)).$$

Thus, the composition of two functions f and g says the following. On input a first apply the function f, then once the output $f(a)$ is computed, apply the function g to the output. For example, let $f : \mathbb{N}^2 \to \mathbb{N}$ and $g : \mathbb{N} \to \mathbb{N}$ be defined as follows:

$$f(n, m) = (n + m)^2 + 2 \quad \text{and} \quad g(k) = k^2,$$

where $(n, m) \in \mathbb{N}^2$ and $k \in \mathbb{N}$. Then the composition $g \circ f$ is the function from $\mathbb{N}^2$ to $\mathbb{N}$ defined as follows:

$$g \circ f(n, m) = ((n + m)^2 + 2)^2.$$

One can apply the composition operation consecutively to several functions. For instance, assume we have three functions

$$f : A \to B \quad \text{and} \quad g : B \to C \quad \text{and} \quad h : C \to D.$$

How would one compose these three functions consecutively? There are two ways to compose these functions. One is to form the composition $(h \circ g) \circ f$. The other is to form the composition $h \circ (g \circ f)$. It turns out these two compositions give the same result. Namely, we have the following property of the composition operation:

Proposition 20.1. *Let* $f : A \to B$, $g : B \to C$, $h : C \to D$ *be functions. Then the following equality, called **associativity**, holds:*

$$(h \circ g) \circ f = h \circ (g \circ f).$$

Proof.

Note that the set A is the set of inputs and D is the set of outputs for both functions $(h \circ g) \circ f$ and $h \circ (g \circ f)$. To prove the proposition, we need to show that for *all* inputs $a \in A$, the equality

$$(h \circ g) \circ f(a) = h \circ (g \circ f)(a)$$

holds true. Indeed, take an $a \in A$. Then we have the following equalities.

$$(h \circ g) \circ f(a) = (h \circ g)(f(a)) = h(g(f(a))) = h \circ g(f(a)) = h \circ (g \circ f)(a).$$

Note that all these equalities above are based on the definition of the composition operation. $\qquad\square$

20.4 Exercises

(1) Consider the partial functions defined below. Find the domain and the range of each of the following partial functions.

    (a) The partial function $f : \mathbb{N} \to \mathbb{N}$ is such that for every natural number n, f outputs k, only when $n = 2k$ where k is a natural number.

    (b) The partial function $f : \mathbb{Z} \to \mathbb{Z}$ is such that for every integer n, if $n \le 0$, then $f(n) = -n$; if $n > 5$, then $f(n) = n$.

    (c) The division function $d : \mathbb{Q}^2 \to \mathbb{Q}$, where $d(q, s) = \frac{q}{s}$ if $s \ne 0$.

    (d) Consider the partial function $f : B \to B$, where B is the set of all binary strings and $f(x) = 1x$ for all $x \in B$. For instance $f(00101) = 100101$.

    (e) Consider the set $A = \{0, 1, \ldots, p - 1\}$, where p is a prime number. The partial function $f : A \to A$ is defined as follows. For all $a \in A$, $f(a) = i + m \pmod{p}$, where m is a fixed number less than p.

(2) Let $f : A \to B$ be a partial function. We say that a function $g : A \to B$ **extends** f if g is total and for every $a \in A$, it must be the case that $g(a) = f(a)$, whenever $f(a)$ is defined. Prove that every partial function $f : A \to B$ can be extended if $B \ne \emptyset$.

(3) Find compositions $f \circ g$ and $g \circ f$ of the following functions:

    (a) $f : \mathbb{N} \to \mathbb{N}$ and $f(n) = n^2$, and $g : \mathbb{N} \to \mathbb{N}$ and $g(n) = n^3 + n^2$, where $n \in \mathbb{N}$.

    (b) $f : \mathbb{Q} \to \mathbb{Q}$ and $f(q) = \frac{q}{q^2+1}$, and $g : \mathbb{Q} \to \mathbb{Q}$ and $g(q) = q^2 - \frac{1}{2}$, where $q \in \mathbb{Q}$.

Programming exercises

For this exercise, all the functions are assumed to be finite.

(1) Design a class to represent finite partial functions.
(2) Write a program that, given a function f, verifies if f is total.
(3) Write a program that, given two functions f and g, determines if $f = g$.
(4) Write a program that, given two functions f and g, outputs $g \circ f$ (if f and g are total).

Lecture 21

Types of functions

The difference between school and life?
In school, you're taught a lesson and then given a test.
In life, you're given a test that teaches you a lesson.
Tom Bodett.

21.1 Surjective functions

In this lecture, we consider only total functions. Recall that a total function is one for which the domain and input sets coincide. The aim of the section is to identify several properties of functions that are useful in the study of functions and their applications.

Let $f : A \to B$ be a function. There is nothing in the definition of functions that guarantees that the range of f coincides with the whole set B. For example, f can send all elements of A onto a given fixed element of B. In this case if B has more than one element, then the range of f does not equal to the set B.

Definition 21.1. We say that the function $f : A \to B$ is a **surjective function** if the range of f is equal to the set B. Sometimes surjective functions are called **onto functions**.

Thus, for the function $f : A \to B$ to be surjective it must be the case that *every* element $b \in B$ has a **pre-image** a, that is, an element $a \in A$ such that $f(a) = b$.

Example 21.1. We now discuss several examples of functions.

(1) Consider the function $S : \mathbb{Z} \to \mathbb{Z}$ given by the rule $S(n) = n + 1$. This function is a surjective function because for every $m \in \mathbb{Z}$ there exists an n such that $S(n) = m$. Indeed, $S(m - 1) = m$ for the given m.

(2) Consider the function $S : \mathbb{N} \to \mathbb{N}$ given by the rule $S(n) = n + 1$. This function is not a surjective function because $0 \in \mathbb{N}$ does not have a pre-image. In other words, for all n from the input set $\mathbb{N}$, we have $S(n) \neq 0$.

(3) Consider the function $f : P(\mathbb{N}) \to \mathbb{N}$ defined as $f(X) = min(X)$ if $X \neq \emptyset$, and $f(X) = 0$ otherwise. This function is surjective. Indeed, take any number $n \in \mathbb{N}$. Clearly $\{n\} \in P(\mathbb{N})$ and $min(\{n\}) = n$. Therefore, $n = f(\{n\})$ by the definition of f. Remember, $P(\mathbb{N})$ is the set of all subsets of $\mathbb{N}$, known as the power set of $\mathbb{N}$.

(4) The addition function $+ : \mathbb{N}^2 \to \mathbb{N}$ is surjective. Indeed, take any number $n \in \mathbb{N}$. Then $n + 0 = n$.

(5) Consider the function $f : \mathbb{Z} \to \mathbb{Z}$ defined as $f(n) = n^2$. This functions is not surjective. For example, no negative number has a pre-image under this function.

21.2 Injective functions

Let $f : A \to B$ be a function. One can view this function as follows. For the purposes of this example, think of the set B as a collection of available codes (such as the integers between 0 and 1000), and think of the set A as a collection of objects that we would like to name by the available codes (such as the collection of employees that work for some company). The function f sets up the coding as follows. The code of the person $p \in A$ is an integer $n \in B$ such that $f(p) = n$. One of the natural requirements put on our coding function f is that we would like no two employees to have the same code. Injective functions express this requirement:

Definition 21.2. We say that the function $f : A \to B$ is an **injective function** if no two distinct elements from A are transformed into the same element of B.

Thus, for a function $f : A \to B$ to be injective f must satisfy the following condition. Whenever $f(x) = f(y)$ then this must imply $x = y$. Here are several examples. We use functions defined in Example 21.1.

Example 21.2.

(1) The function $S : \mathbb{Z} \to \mathbb{Z}$ is an injective function. Indeed, if $S(x) = S(y)$ then $x + 1 = y + 1$ which implies $x = y$.

(2) The function $S : \mathbb{N} \to \mathbb{N}$ is also an injective functions by the same reason as above.

(3) The function $f : P(\mathbb{N}) \to \mathbb{N}$ is not an injective function. Indeed, consider for example the sets $\{n\}$ and $\{n, n + 1\}$. These two distinct elements of $P(\mathbb{N})$ are both transformed into n.

(4) The addition function $+ : \mathbb{N}^2 \to \mathbb{N}$ is not an injective function. Indeed, two distinct pairs, for example, $(n, 0)$ and $(0, n)$ are transformed into the same element n under the addition.

(5) Consider the function $f : \mathbb{Z} \to \mathbb{Z}$ defined as $f(n) = n^2$. This functions is not an injective function. For example, for any integer n, we have $f(n) = f(-n)$.

(6) Let B be the set of all binary strings over the alphabet $\{0, 1\}$. Consider the function $f : B \to B$ such that for any string $x = a_0 \ldots a_n$, we have $f(x) = a_1 a_2 \ldots a_n a_0$. Thus, f *rotates* any input x and outputs the result. This function is injective. Indeed, assume $f(x) = f(y)$ and say $x = a_0 \ldots a_n$ and $y = b_0 \ldots b_n$. Then by the definition of f and the equality $f(x) = f(y)$, we have:

$$a_1 a_2 \ldots a_n a_0 = f(x) = f(y) = b_1 b_2 \ldots b_n b_0.$$

Hence, $a_1 = b_1$, $a_2 = b_2$, ..., $a_n = b_n$, and $a_0 = b_0$. Therefore $x = y$.

21.3 Bijective functions

Let $f : A \to B$ be a function. As in the previous section, let us view this function as a function that encodes the objects in A by codes in B. For the function f to be a coding function there are two natural requirements. One is that no two objects in A have the same code in B. Ideally, we would like every code in B to represent some object in A. Clearly, the first requirement is that f is injective. The second requirement is that f is surjective. We put these two requirements into the following definition:

Definition 21.3. Let $f : A \to B$ be a function. We say that the function f is a **bijective function** if it is both surjective and injective. Bijective functions are also called **one-to-one** functions.

Thus, a bijective function from A to B tells us that there is a one-to-one correspondence between the elements of set A and the elements of set B. For instance, the function $S : \mathbb{Z} \to \mathbb{Z}$ in the examples above is bijective, whereas $S : \mathbb{N} \to \mathbb{N}$ is not. Here is another example:

Example 21.3. Let $f : \mathbb{Z} \to \mathbb{Z}$ be a function defined as: $f(n) = (-1)^n \cdot 3 + n$. We show that this function is a bijective function by proving that f is both surjective and injective.

(1) We first show that f is an injective function. Take two distinct integers n and m. We want to prove that $f(n) \neq f(m)$. For the proof, there are three cases to consider.

   Case 1. n is even and m is odd. In this case $f(n) = 3 + n$ and $f(m) = m - 3$. Hence $f(n)$ is odd and $f(m)$ is even. Therefore $f(n) \neq f(m)$. Note that the case when n is odd and m is even is treated similarly.

   Case 2. Both n, m are even. In this case $f(n) = n+3$ and $f(m) = m+3$. Therefore, we have $f(n) \neq f(m)$ since $n \neq m$ by our assumption.

   Case 3. Both n, m are odd. In this case $f(n) = n-3$ and $f(m) = m-3$. Therefore, we have $f(n) \neq f(m)$ since $n \neq m$.

(2) We now show that f is a surjective function. Take any integer n. If n is odd then $f(n - 3) = n$. If n is even then $f(n + 3) = n$. Thus, f is onto.

We have proved that f is a bijective function.

21.4   Transition functions

In this section, we want to use functions as a tool to model computer programs. A typical behavior of a program can be described as follows. At any given time during execution, the program is in some state. The state is usually determined by the current values of all the variables in the program. For example, a state of a robot control program would contain information about the position of the robot and its surroundings. Given a state and an input signal, the program either stays in the same state or makes a transition to another state. The input signal can be received from the user, the surroundings, another program, or an instruction from the program itself. Thus, the subsequent state of the program is determined by the current state and the input signal. The move of the program from its current state to the subsequent state is called a transition of the program. The program includes a finite number of variables and the domains of the variables are all finite. Therefore, we can always assume that the number of states of the program is finite. We now formally model the reasoning above.

Definition 21.4. A **transition function** is a function $T : S \times \Sigma \to S$ where:

(1) The set S is the **set of states** of T,
(2) The set Σ is the **set of input symbols** of T.

The set Σ is also referred to as the **alphabet** of T. The elements of Σ are **input symbols** or **letters** of the alphabet Σ.

Given a transition function $T : S \times \Sigma \to S$, we can view of the set S as the state space of a program and T as its the transitions in the state space S. For a state $s \in S$ and an input signal $\sigma \in \Sigma$, if $T(s, \sigma) = q$ then we interpret this as follows. Whenever the program is in the state s and reads the input signal σ then the program transitions into state q.

An example of a transition function is the transition of states when we add numbers (in binary). In this case, there are two states. One state is the carry state, let us denote it by 1; and the other state is the non-carry state, let us denote it by 0. The addition is performed bit by bit. Therefore, the inputs to these two states are of the form

$$\begin{pmatrix} 1 \\ 1 \end{pmatrix}, \begin{pmatrix} 1 \\ 0 \end{pmatrix}, \begin{pmatrix} 0 \\ 1 \end{pmatrix}, \begin{pmatrix} 0 \\ 0 \end{pmatrix}.$$

The table below represents the transition function for addition. For example, when the state is the carry state (1) and the input is $\begin{pmatrix} 1 \\ 0 \end{pmatrix}$, then the next state is still the carry state (1).

	$\begin{pmatrix} 1 \\ 1 \end{pmatrix}$	$\begin{pmatrix} 1 \\ 0 \end{pmatrix}$	$\begin{pmatrix} 0 \\ 1 \end{pmatrix}$	$\begin{pmatrix} 0 \\ 0 \end{pmatrix}$
0	1	0	0	0
1	1	1	1	0

We present other examples of transition functions, T_1 and T_2, given in the tables below, respectively. For instance, the states of the transition function T_1 are 1, 2, and 3. The input symbols are a, b and c. For example, when the transition system is in state 3 and reads input symbol c, the function makes a transition to state 2.

T_1	a	b	c
1	1	2	2
2	2	3	2
3	3	1	2

T_2	a	b	c
1	3	2	1
2	1	2	4
3	2	1	4
4	2	3	3

Transition functions $T : S \times \Sigma \to S$ can be represented using directed graphs as follows. First, represent states s from S as vertices of the graph. Second, put an edge from state s to state q if there is an input σ such that $T(s, \sigma) = q$. Finally, label this edge with the symbol σ. Such representations of transition functions are called **transition diagrams**. For example, the transition diagrams for the addition function on binary numbers and the transition function T_2 are shown in Figure 21.1.

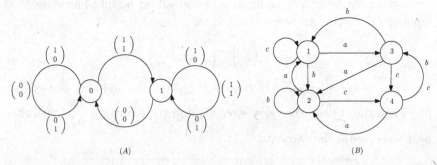

Fig. 21.1: (A): The diagram for binary addition. (B): The diagram for T_2.

There is an important operation called the product operation. When applied to two transition functions T_1 and T_2, it produces a new transition function called the **product of T_1 and T_2**. Intuitively, the product of T_1 and T_2 is a transition function that simulates both T_1 and T_2 simultaneously. To explain this operation, assume that we are given two transition functions

$$T_1 : S_1 \times \Sigma \to S_1 \quad \text{and} \quad T_2 : S_2 \times \Sigma \to S_2.$$

The product of T_1 and T_2 is the transition function T defined as follows:

(1) The set S of states of T is the Cartesian product $S_1 \times S_2$.
(2) The set of input symbols of T is Σ.
(3) The function T is defined by the following rule. For all $(s_1, s_2) \in S$ and $\sigma \in \Sigma$:

$$T((s_1, s_2), \sigma) = (T_1(s_1, \sigma), T_2(s_2, \sigma)).$$

From this definition we see that the product T of T_1 and T_2 makes transitions as follows. If T is in a state (s_1, s_2) and reads an input signal σ, then the first component of T simulates T_1, and the second component of T simulates T_2. As an example, the product of the transition tables T_1 and T_2 given above is presented below.

	a	b	c
(1,1)	(1,3)	(2,2)	(2,1)
(1,2)	(1,1)	(2,2)	(2,4)
(1,3)	(1,2)	(2,1)	(2,4)
(1,4)	(1,2)	(2,3)	(2,3)
(2,1)	(2,3)	(3,2)	(2,1)
(2,2)	(2,1)	(3,2)	(2,4)
(2,3)	(2,2)	(3,1)	(2,4)
(2,4)	(2,2)	(3,3)	(2,3)
(3,1)	(3,3)	(1,2)	(2,1)
(3,2)	(3,1)	(1,2)	(2,4)
(3,3)	(3,2)	(1,1)	(2,4)
(3,4)	(3,2)	(1,3)	(2,3)

Table 21.1: The product of T_1 and T_2

21.5 Exercises

(1) Let B be the set of all binary strings over the alphabet $\{0, 1\}$. Consider the function $f : B \to B$ such that for any string x, the value $f(x)$ is

obtained by replacing all 0's in x by 1's, and all 1's in x by 0's. Is the function bijective? Explain your answer.

(2) Consider the function $f : \mathbb{Z}^2 \to \mathbb{Z}^2$ defined as follows. For $(x, y) \in \mathbb{Z}^2$: if both x and y are even or both x and y are odd then $f(x, y) = (x, y+1)$; otherwise $f(x, y) = (x + 1, y)$. Prove that f is a bijective function.

(3) Let A and B be two finite sets. Show that there exists a bijective function $f : A \to B$ if and only if A and B have the same number of elements.

(4) Let $f : \mathbb{Z} \to \mathbb{Z}$ be the function defined as: $f(n) = (-1)^n + n$. Prove that f is a bijective function.

(5) Let A be a set. Consider the set of *all* functions from A into $\{0, 1\}$. This set is denoted by 2^A. Prove that there exists a bijective function from $P(A)$ to 2^A.

(6) Give an example of a bijective function f from $\mathbb{N}$ to $\mathbb{Z}$.

(7) Give an example of a bijective function from $\mathbb{N}$ to $\mathbb{N}^2$. (*Hint:* Draw the grid for $\mathbb{N}^2$ and start enumerating the grid elements in a systematic way. This produces a desired function.)

(8) Prove that the composition of onto functions is onto.

(9) Prove that the composition of injective functions is injective.

(10) Draw all possible transition diagrams of the transition functions $T : S \times \{a, b\} \to S$, where $S = \{0, 1\}$.

(11) Let E be an equivalence relation on $\mathbb{N}$. The **factor set** defined by E is the set of all E-equivalence classes. Denote the factor set by $\mathbb{N}/E$. Call a number $n \in \mathbb{N}$ a **representative** if n is the minimal element in the equivalence class containing n. Prove that there exists a bijection between the factor set and the set of all representatives.

Programming exercises

(1) Write a program that, given a function f, determines if f is surjective.

(2) Write a program that, given a function f, determines if f is injective.

(3) Write a program that, given a function f, determines if f is bijective.

(4) Write a program that, given two transition functions T_1 and T_2, outputs their product.

Lecture 22

Syntax of propositional logic

Logic is the beginning of wisdom, not the end.
Leonard Nemoy.

22.1   Propositions and formulas

In modern computer science, software engineering and mathematics logic plays an important role. In fact, logic has been called the calculus of computer science. Logic is used in computability theory (a mathematical study of what can and what cannot be computed), complexity theory (a study of what can and what cannot be computed efficiently), databases, information retrieval, design verification, programming languages, circuit design, AI, software engineering, security, and many more areas of computer and other sciences.

The historical roots of logic go back to studies in philosophy and mathematics. Our goal in this lecture is to develop the syntax of propositional logic, and to give some proofs that use our knowledge about induction.

A *proposition* is a simple statement that can be evaluated to true or false. A good example of a proposition is the following. Assume that we have a program that uses a variable v ranging over its domain D. Let us take a subset $A \subseteq D$. For example, D can be the set of integers, and A can be the set of even integers. Consider, the following proposition S:

v belongs to set A.

Let us assume that we run the program on some inputs. At any given stage of the run, the proposition S takes either the value true or false. For example, in the Euclidean algorithm, see Lecture 4, there is a variable r whose value changes over the domain of integers throughout the execution

213

of the program. For the given set A of even integers, the statement

$$r \ belongs \ to \ set \ A$$

can either be true or false. Here are more examples of propositions that can be evaluated to true of false: (1) The roses are red. (2) The value of an integer variable is odd. (3) Aristotle is a philosopher. (4) The result of $4 + 2$ is 8.

For this lecture, we abstract away from the actual meanings of the propositions. Instead, we are interested in how we can construct more complicated statements from simple propositions. To be more formal we start with the following definition.

Definition 22.1. A **proposition** is a simple sentence whose value is either **true** or **false**. We denote propositions by letters p, q, etc. The collection of all propositions is denoted by $PROP$.

We can view the set $PROP$ of propositions as the set of variables that can be evaluated to either **true** or **false**. From a programming point of view, propositions are simply Boolean variables. We often identify 0 with **false** and 1 with **true**.

We build complicated statements from propositions using the following **connectives**:

(1) The **and** connective denoted by $\wedge$.
(2) The **or** connective denoted by $\vee$.
(3) The **if ... then ...** connective denoted by $\rightarrow$.
(4) The **not** connective denoted by $\neg$.

The first three connectives are applied to a pair of statements, and are thus called *binary connectives*. The last connective is applied to a single statement, and is called a *unary connective*. Here are some examples of statements built from propositions and that use the connectives above:

(1) The roses are red or the roses are yellow.
(2) If a student asks a question then the instructor answers.
(3) The value of an integer variable is even and positive.
(4) Aristotle is a man and he is a philosopher.
(5) It is not the case that $4 + 2 = 8$.

Thus, as seen from these examples, the connectives above can be used to obtain more complicated sentences. As an example, consider the statement

If a student asks a question then the instructor answers.

If we denote the sentence *"a student asks a question"* with p, and the sentence *"the instructor answers"* with q then the original statement can be written as $(p \to q)$. Similarly, the statement

It is not the case that $4 + 2 = 8$

can be written as $\neg(s)$, where s denotes the statement $4 + 2 = 8$. These more complicated sentences built from the propositions are called formulas.

We ignore the meaning of these formulas and give a formal definition as how we build them. Our definition is an inductive definition.

Definition 22.2. The set of **formulas** of propositional logic, denoted by $FORM$, is defined inductively as follows:

Base case: Every proposition $p \in PROP$ is a formula.

Inductive step: Let ϕ and ψ be formulas. Then each of the following is a formula: $(\phi \land \psi)$, $(\phi \lor \psi)$, $(\phi \to \psi)$, and $\neg(\phi)$.

Examples of formulas are

$$p, \ q, \ (p \land q), \ \neg(p), \ \neg((p \land q)), \ (q \to \neg((p \land q))).$$

The following are not formulas:

$$\neg p, \ p \land q, \ (p \land q) \lor s, \ (q \to s) \land (p \lor t)), \ \neg(p \land q).$$

Formulas that are not propositions are called **compound formulas**. Every compound formula must start with either the left parenthesis (or the negation symbol $\neg$, and must end with the right parenthesis). We denote formulas with the letters of the Greek alphabet such as

$$\alpha, \ \beta, \ \gamma, \ \phi, \text{ and } \psi.$$

We comment that the connectives $\land$, $\lor$, $\neg$, and $\to$ can be viewed as constructors that build formulas. We emphasize that formulas are just special type of strings built from propositions, connectives and parentheses. However, not all strings are formulas. Only the strings that can be constructed as defined above are formulas. The next section explains how formulas can be viewed as labeled trees. This will give a more visual and intuitive representation of formulas.

22.2   Formation trees for formulas

Our initial goal is to study some of the *syntactic properties* of formulas. Clearly, every formula is simply a finite string over the alphabet $\{\vee, \wedge, \rightarrow, \neg, (,)\} \cup PROP$. Therefore, we can talk about the equality of formulas as strings. For instance, the formulas $(p \wedge q)$ and $(q \wedge p)$ are not equal.

Definition 22.3. We say that two formulas ϕ and ψ are **equal**, denoted as $\phi = \psi$, if they are equal as strings.

Each formula ϕ can be represented as a labeled tree T_ϕ, which is called the **formation tree**. We construct formation trees of formulas by induction as follows:

Definition 22.4.
Basis: If ϕ is a proposition p then T_ϕ is the tree consisting of only the root, where the root is labelled by p.
Inductive Step: Assume that we have constructed labeled trees T_α and T_β for formulas α and β. Assume that $\phi = (\alpha \vee \beta)$. The tree T_ϕ is constructed as follows:

(1) Create a new root r and label it with $\vee$.
(2) Declare the left subtree of the root r to be T_α.
(3) Declare the right subtree of the root r to be T_β.
(4) Preserve all the labels of T_α and T_β.

The labeled trees for $(\alpha \wedge \beta)$ and $(\alpha \rightarrow \beta)$ are constructed in a similar way. The tree $T_{\neg(\alpha)}$ for $\neg(\alpha)$ is constructed as follows:

(1) Create a new root r and label it with $\neg$.
(2) Declare the root of T_α to be the only child of the new root r.
(3) Preserve all the labels of T_α.

For instance, the formation tree representing the formula
$$(\ (\ \neg(p) \vee (\ q \rightarrow (\neg(p) \wedge s)\ )\ )\ ) \vee (\ \neg((p \vee t)) \rightarrow (s \wedge \neg(t))\ )\ )$$
is given in Figure 22.1. Now our goal is to show that every formula ϕ has a *unique* tree T_ϕ that represents it. In order to show this, we need to prove several syntactic properties of formulas. Since formulas are defined by induction, most of their properties are also proved by induction.

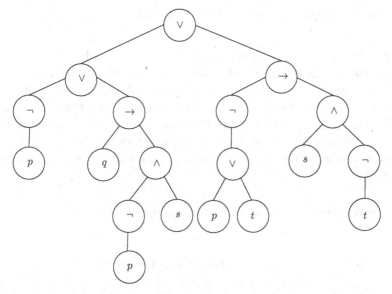

Fig. 22.1: A formation tree for ((¬(p)∨(q → (¬(p)∧s))) ∨ (¬((p∨t)) → (s ∧ ¬(t))))

22.3 The unique readability theorem

Here we prove that every formula of propositional logic is uniquely written. This is the most important syntactic property of formulas. The importance of this will be seen in the next two lectures when we start to give meanings to the formulas through their truth values. All the proofs below are inductive proofs owing to the inductive definition of formulas. We start with the following lemma.

Lemma 22.1. *Assume that ϕ and ψ are formulas. If the formula ψ starts with ϕ then $\phi = \psi$.*

Proof.
We prove the lemma by induction on the length of ϕ. If the length of ϕ is 1 then ϕ must be a proposition. Moreover, since ψ starts with ϕ, it must be the case that $\phi = \psi$.

Assume now that ϕ starts with ¬. Our inductive hypothesis is that the lemma is true for all formulas of a smaller length than ϕ. Since ϕ starts with ¬, the formula ϕ is of the form ¬(ϕ') where ϕ' is a formula. Since ψ starts with ϕ, it must be the case that ψ is of the form ¬(ψ') where ψ' is a formula. Therefore, ψ' starts with ϕ'. By induction hypothesis $\phi' = \psi'$, and we conclude that $\phi = \psi$.

Assume now that ϕ is of the form $(\phi_0 \vee \phi_1)$. Our inductive hypothesis is that the lemma is true for all formulas of a smaller length than ϕ. The formula ψ must be of the form $(\psi_0 \tau \psi_1)$, where τ is one of the symbols $\vee$, $\wedge$ or $\rightarrow$. Since ϕ is the start of ψ, it must be the case that either ϕ_0 starts with ψ_0 or ψ_0 starts with ϕ_0. In either case, by the inductive assumption $\phi_0 = \psi_0$. Hence $\tau = \vee$ and $\phi_1 = \psi_1$. We conclude that $\phi = \psi$. The case when ϕ is either of the form $(\phi_0 \wedge \phi_1)$ or of the form $(\phi_0 \rightarrow \phi_1)$ is treated similarly. $\square$

Now we prove an important syntactic property of formulas.

Theorem 22.1 (The unique readability theorem). *Every formula α is either a proposition or can be uniquely written in one of the following forms:* $(\phi \vee \psi)$, $(\phi \wedge \psi)$, $(\phi \rightarrow \psi)$, $\neg(\phi)$.

Proof.
We prove the theorem by induction on the length of α. If α is a proposition then we are done. Our induction hypothesis is now that the theorem is true for all formulas of length less than α.

Assume that α starts with a $\neg$. Then α is of the form $\neg(\phi_1)$. If α can also be written in the form $\neg(\phi_2)$ then it must be the case either ϕ_1 starts with ϕ_2 or ϕ_2 starts with ϕ_1. By the lemma above, we then have $\phi_1 = \phi_2$.

Assume that α is of the form $(\phi_1 \vee \phi_2)$. Suppose that α can also be written as $(\psi_1 \tau \psi_2)$, where $\tau \in \{\vee, \wedge, \rightarrow\}$. Then it must be the case that either ϕ_1 starts with ψ_1 or ψ_1 starts with ϕ_1. In either case, by the lemma above, $\phi_1 = \psi_1$. Therefore, $\tau = \vee$, and hence $\phi_2 = \psi_2$. The case when ϕ is of the form $(\phi_0 \wedge \phi_1)$ or of the form $(\phi_0 \rightarrow \phi_1)$ is treated similarly. $\square$

This theorem guarantees that each formula ϕ has a unique tree T_ϕ representing it. Thus, we have the following corollary:

Corollary 22.1. *Every formula ϕ of propositional logic has a unique formation tree T_ϕ associated with ϕ.*

22.4 Exercises

(1) Write all formulas of length at most 5 that use propositions p and q.

(2) Design an algorithm that given a string over the alphabet

$$\{p_1, \ldots, p_n,), (, \wedge, \vee, \neg\}$$

detects if the string is a formula.

(3) Let ϕ be formula. Prove that with every occurrence of (in ϕ there is a sub-formula of ϕ uniquely associated with (. For example, in the formula $\phi = ((p \wedge q) \vee (q \to \neg(p \wedge q)))$ the sub-formula associated with the first occurrence of (is the ϕ itself, the sub-formula associated with the second occurrence of (is $(p \wedge q)$, and the formula associated with the third occurrence of (is $(q \to \neg(p \wedge q))$.

(4) Draw formation trees for the following formulas:

   (a) $(p \wedge q)$,

   (b) $\neg((p \wedge q))$,

   (c) $(q \to \neg((p \wedge q)))$

   (d) $(r \to \neg((q \to \neg((p \wedge q)))))$.

Programming exercises

(1) Write a program that checks if an input string is a formula built from propositions $\{p_0, \ldots, p_n\}$.

(2) Write a program that given a formula ϕ and an occurrence (in the formula outputs a sub-formula uniquely associated with the occurrence (.

(3) Write a program that given two formulas α and β determines if β is a sub-formula of α.

(4) Write a program that given a formula ϕ builds the formation tree T_ϕ.

Lecture 23

Semantics of propositional logic

The truth is incontrovertible. Malice may attack it,
ignorance may deride it, but in the end, there it is.
Winston Churchill.

23.1 Truth assignments

As we mentioned in the previous lecture, every proposition can be viewed as a Boolean variable that takes value **true** or **false**. Therefore, we can assign truth assignments to propositions.

Definition 23.1. A **truth assignment** is a function

$$A : PROP \to \{\text{true}, \text{false}\}.$$

Thus, a truth assignment A assigns to each proposition $p \in PROP$ a unique truth value $A(p) \in \{\text{true}, \text{false}\}$. There are exactly 2^n distinct truth assignments of the propositions $p_1, \ldots, p_n$.

Example 23.1. For two propositions, say p_1 and p_2, there are four different truth assignments possible. We call them, A_1, A_2, A_3, and A_4, which operate as follows:

(1) $A_1(p_1) = \text{true}$, $A_1(p_2) = \text{true}$,
(2) $A_2(p_1) = \text{true}$, $A_2(p_2) = \text{false}$,
(3) $A_3(p_1) = \text{false}$, $A_3(p_2) = \text{true}$, and
(4) $A_4(p_1) = \text{false}$, $A_4(p_2) = \text{false}$.

Suppose that we are given a truth assignment A. Note that A gives truth values to propositions but not to compound formulas. We would like to use A to compute truth values to all formulas. To do this, we need to interpret the connectives $\lor$, $\land$, $\to$, and $\neg$ as functions that output **true** or **false** Boolean values.

We start with the interpretation of the negation symbol $\neg$ in the table below. This negation table states an obvious fact that the negation of **true** is **false**, and the negation of **false** is **true**.

$\neg(\textbf{true}) = \textbf{false}$
$\neg(\textbf{false}) = \textbf{true}$

The interpretation of the connective $\lor$ is presented in the next table. This table gives a natural interpretation of the *or* connective $\lor$. For example, the truth value of (**true** $\lor$ **false**) should clearly be **true**. Note that the interpretation of $\lor$ is *not* exclusive but rather inclusive.

true $\lor$ **true** = **true**
true $\lor$ **false** = **true**
false $\lor$ **true** = **true**
false $\lor$ **false** = **false**

The following table interprets the *and* connective $\land$, also in an obvious way. The only time that $\land$ gives a **true** value is when both its arguments have a **true** value.

true $\land$ **true** = **true**
true $\land$ **false** = **false**
false $\land$ **true** = **false**
false $\land$ **false** = **false**

Finally, the table below interprets the *if ... then ...* connective $\to$. The evaluation of (**false** $\to$ **true**) as being **true** typically causes some controversy. One explanation for this evaluation, from a mathematics point of view, is this. Consider a statement of the form

if H then C

that we already discussed in the first lectures. To prove that this statement is true we assume that H is true and then prove that C is true. If H is not true then there is nothing to prove, and hence by default one evaluates the statement as true. Therefore, it makes sense to evaluate (**false** → **true**) as **true**.

true → **true** = **true**
true → **false** = **false**
false → **true** = **true**
false → **false** = **true**

Let $A : PROP \rightarrow \{T, F\}$ be a truth assignment. Consider a formula ϕ. Using the interpretations of the $\neg$, $\vee$, $\wedge$ and $\rightarrow$ as explained in the tables above, we can *lift* the assignment A to evaluate the formula ϕ. As an example, consider the formula $\phi = ((p_1 \rightarrow p_2) \vee \neg(p_1))$. As explained in Example 23.1, there are four possible truth assignments A_1, A_2, A_3 and A_4 of the propositions p_1 and p_2. Let us take A_2 for example, which says $A_2(p_1) = $ **true**, $A_2(p_2) = $ **false**. Hence, the truth value of $(p_1 \rightarrow p_2)$ is **false** and the truth value of $\neg(p_1)$ is also **false**. The value of (**false** $\vee$ **false**) is **false**. Hence, the formula ϕ is evaluated to **false** under A_2.

The above reasoning suggests that, for any given assignment A, we can *lift* A and give the truth value, denoted by $V(\phi)$, to every formula ϕ using the assignment A.

Definition 23.2.
Base case: If formula ϕ is a proposition p then set $V(\phi) = A(p)$.

Inductive step: Assume that the formula ϕ is of the form $\neg(\psi)$ and V has been defined for all sub-formulas of ϕ. In this case, set

$$V(\phi) = \neg V(\psi).$$

Assume that the formula ϕ is of the form $(\phi_1 \vee \phi_2)$. By the inductive assumption, V has been defined on ϕ_1 and ϕ_2. Set

$$V(\phi) = V(\phi_1) \vee V(\phi_2).$$

Similarly, we evaluate $(\phi_1 \wedge \phi_2)$ and $(\phi_1 \rightarrow \phi_2)$ using the interpretations of the connectives $\wedge$ and $\rightarrow$ from the corresponding tables above. Note that V depends on A and therefore we sometimes write V as V_A to indicate the dependence of V on A.

In this definition, the connectives $\neg$, $\vee$, $\wedge$ and $\rightarrow$ all use the truth table presented earlier.

Here is an important point. The definition of V_A uses the unique readability theorem. The theorem is used in the inductive step. Indeed, say ϕ is of the form $(\phi_1 \vee \phi_2)$. The theorem states that ϕ can be written uniquely as $(\phi_1 \vee \phi_2)$. Thus, to evaluate ϕ it suffices to evaluate both ϕ_1 and ϕ_2. The unique readability theorem guarantees that this is a correct evaluation.

We call V_A the **truth valuation** of formulas defined by the assignment A. Clearly the truth valuation V_A is a mapping from the set of formulas $FORM$ to $\{\textbf{true}, \textbf{false}\}$. The following theorem collects the basic properties of the evaluation V_A. The proofs of all of these properties simply follow from the definition of V_A we have just given above.

Theorem 23.1. *Let* $A : PROP \to \{\textbf{true}, \textbf{false}\}$ *be a truth assignment. Then for all formulas the truth valuation* $V_A : FORM \to \{\textbf{true}, \textbf{false}\}$ *satisfies the following properties:*

(1) $V_A(\neg(\phi)) = \neg V_A(\phi)$.
(2) $V_A(\phi \vee \psi) = V_A(\phi) \vee V_A(\psi)$.
(3) $V_A(\phi \wedge \psi) = V_A(\phi) \wedge V_A(\psi)$.
(4) $V_A(\phi \to \psi) = V_A(\phi) \to V_A(\psi)$.$\square$

We stress that in this theorem the connectives $\wedge$, $\vee$, $\to$, and $\neg$ are used in two distinct ways. One is that they are used as symbols (on the left side of the equations), and the other is that they are used as interpretations (on the right side of the equation) presented in the tables.

Consider a formula ϕ. It is built from a finite number of propositions. Therefore, in order to indicate all the propositions from which ϕ is built, we define:

$$PROP(\phi) = \{p \mid \text{proposition } p \text{ occurs in } \phi\}.$$

For example, for

$$\phi = (\ (\ \neg(p) \vee (\ q \to (\neg(p) \wedge s)\ )\ ) \vee (\ (p \vee t) \to (s \wedge \neg(t))\ )\ )$$

we have

$$PROP(\phi) = \{p, q, s, t\}.$$

Clearly $PROP(\phi)$ is a finite set for every formula ϕ.

Definition 23.3. The truth assignments A_1 and A_2 **agree** on set $PROP(\phi)$ if $A_1(p) = A_2(p)$ for all $p \in PROP(\phi)$.

The next theorem shows that the propositions that do not occur in ϕ have no impact on the truth value of ϕ. In other words, the propositions that are relevant in evaluating ϕ are just those that belong to $PROP(\phi)$.

Theorem 23.2 (The relevance theorem). *If truth assignments A_1 and A_2 agree on $PROP(\phi)$ then $V_{A_1}(\phi) = V_{A_2}(\phi)$.*

Proof.
The proof of the theorem is based on the inductive definitions of V_{A_1} and V_{A_2}.

Basis. Assume ϕ is a proposition, say p. Then, since A_1 and A_2 agree on propositions we have:

$$V_{A_1}(\phi) = A_1(p) = A_2(p) = V_{A_2}(\phi).$$

Induction step: Assume that the formula ϕ is of the form $\neg(\psi)$. By the induction hypothesis, we have $V_{A_1}(\psi) = V_{A_2}(\psi)$. Therefore, by the inductive definitions of V_{A_1} and V_{A_2} we have:

$$V_{A_1}(\phi) = \neg V_{A_1}(\psi) = \neg V_{A_2}(\psi) = V_{A_2}(\phi).$$

Assume that the formula ϕ is of the form $(\phi_1 \tau \phi_2)$, where $\tau \in \{\vee, \wedge, \rightarrow\}$. By the inductive assumption, we have $V_{A_1}(\phi_1) = V_{A_2}(\phi_1)$ and $V_{A_1}(\phi_2) = V_{A_2}(\phi_2)$. Hence we have:

$$V_{A_1}(\phi) = V_{A_1}(\phi_1) \ \tau \ V_{A_1}(\phi_2) = V_{A_2}(\phi_1) \ \tau \ V_{A_2}(\phi_2) = V_{A_2}(\phi). \qquad \square$$

23.2 Logical equivalence

Given a formula ϕ, we can evaluate it under *all* possible truth assignments. For example, consider the formula ϕ:

$$\phi = ((p \vee \neg(q)) \rightarrow (s \wedge q)).$$

Its propositions are p, q, and s. There are exactly 8 possible assignments to these propositions. Each possible assignment gives rise to a truth evaluation of ϕ. We show all these assignments in the table below. In the table, we add columns labeled by sub-formulas, to make our calculations easier. These types of tables are called **truth tables**.

p	q	s	$(p \vee \neg(q))$	$(s \wedge q)$	ϕ
T	T	T	T	T	T
T	T	F	T	F	F
T	F	T	T	F	F
T	F	F	T	F	F
F	T	T	F	T	T
F	T	F	F	F	T
F	F	T	T	F	F
F	F	F	T	F	F

One of the important concepts in propositional logic is the notion of equivalence. The idea is that two formulas α and β that are syntactically different might actually be semantically the same. For example, consider the formulas $(p \to q)$ and $(\neg(p) \vee q)$. Under all possible truth assignments these two formulas are evaluated to the same truth value. We formalize this observation in the next definition:

Definition 23.4. We say that formulas ϕ and ψ are **equivalent** if $V_A(\phi) = V_A(\psi)$ for *all* truth assignments A.

As an example, the following pairs of formulas are all equivalent (see Exercise 3):

(1) ϕ and $(\phi \vee \phi)$.
(2) $(\phi \wedge (\alpha \vee \beta))$ and $((\phi \wedge \alpha) \vee (\phi \wedge \beta))$.
(3) ϕ and $\neg\neg\phi$.
(4) $(\phi \vee (\alpha \wedge \beta))$ and $((\phi \vee \alpha) \wedge (\phi \vee \beta))$.
(5) $\neg(\phi \vee \psi)$ and $(\neg\phi \wedge \neg\psi)$.
(6) $\neg(\phi \wedge \psi)$ and $(\neg\phi \vee \neg\psi)$.
(7) $(\alpha \vee (\beta \vee \gamma))$ and $((\alpha \vee \beta) \vee \gamma)$.
(8) $(\alpha \wedge (\beta \wedge \gamma))$ and $((\alpha \wedge \beta) \wedge \gamma)$.

In this list of examples and henceforth, we abuse notation if it does not cause confusion. For instance, we can write a conjunction of formulas as follows:

$$(\phi_1 \wedge \phi_2 \wedge \ldots \wedge \phi_n).$$

In this notation, we drop all the parentheses inside of the formula and keep the leftmost and the rightmost parentheses. Similarly, if all connectives of the formula are disjunctions then we can write this formula as

$$(\psi_1 \vee \psi_2 \vee \ldots \vee \psi_k).$$

We can also remove the parentheses in formulas $\neg(\phi)$ by writing them in a simpler form as $\neg\phi$.

Finally, we note that formulas ϕ and ψ may be equivalent even if one contains propositions that the other does not. For example, $(p \wedge (s \to s))$ and $(p \wedge (q \vee \neg q))$ are equivalent.

23.3   Exercises

(1) Consider the formula

$$\phi = (\ (\ \neg(p_1) \vee (\ p_2 \to (\neg(p_1) \wedge p_3)\ )\ )\ ) \vee (\ (p_1 \vee p_3) \to (p_3 \wedge \neg(p_2))\ )\ )\ )$$

Write down all truth assignments of the propositions p_1, p_2, and p_3. For each truth assignment A, find the truth value $V_A(\phi)$ of the formula ϕ.

(2) Prove that there are exactly 2^n truth assignments to propositions p_1, $\ldots$, p_n.

(3) Prove that the following pairs of formulas are equivalent:

   (a) $(\phi \to \psi)$ and $(\neg\phi \vee \psi)$.
   (b) $(\phi \wedge (\alpha \vee \beta))$ and $((\phi \wedge \alpha) \vee (\phi \wedge \beta))$.
   (c) ϕ and $\neg\neg\phi$.
   (d) $\neg(\phi \vee \psi)$ and $(\neg\phi \wedge \neg\psi)$.
   (e) $\neg(\phi \wedge \psi)$ and $(\neg\phi \vee \neg\psi)$.
   (f) $(\alpha \wedge (\beta \wedge \gamma))$ and $((\alpha \wedge \beta) \wedge \gamma)$.
   (g) Prove that for all truth assignment A and formulas ϕ and $(\phi \to \psi)$ if $V_A(\phi) = $ **true** and $V_A(\phi \to \psi) = $ **true** then $V_A(\psi) = $ **true**.

Programming exercises

(1) Write a program that, given a truth assignment A of propositions p_1, $\ldots$, p_n and a formula ϕ, evaluates ϕ under the truth assignment A.

(2) Write a program that given a formula ϕ outputs all propositions from $Prop(\phi)$.

(3) Write a program that determines whether two formulas ϕ and ψ are logically equivalent. Beware that this program may take a long time to compute for formulas with many propositions.

Lecture 24

Normal forms and the SAT problem

> *Some mathematician has said pleasure lies*
> *not in discovering truth, but in seeking it.*
> Lev Tolstoy.

24.1 Truth tables and normal forms

In the previous lecture we gave an example of a truth table for the formula

$$\phi = ((p \vee \neg q) \to (s \wedge q)).$$

Naturally, one can build a truth table for any given formula of the propositional logic. Indeed, let ϕ be a formula such that

$$PROP(\phi) = \{q_1, \ldots, q_n\}.$$

We build a table with $n + 1$ columns labeled with q_1, q_2, ..., q_n, and ϕ, respectively. Each row of the table represents a truth assignment given to the propositions q_1, ..., q_n and the truth value of ϕ under that assignment. There are exactly 2^n rows. Such a table is called the **truth table** for formula ϕ (see the table below). Let us denote the assignment that corresponds to the i-th row by A_i. Each such assignment evaluates the original formula ϕ. Let X_i be the truth value of ϕ under the assignment A_i. The truth table for ϕ conveys all this information.

q_1	q_2	... ...	q_{n-1}	q_n	ϕ
T	T	... ...	T	T	X_1
F	T	... ...	T	F	X_2
...	...	... ...	...	...	...
...	...	... ...	...	...	...
F	F	... ...	F	F	X_{2^n}

Recall from the previous lecture that two formulas ϕ and ψ are **equivalent** if for all truth assignments A the formulas ϕ and ψ are evaluated to the same truth value, that is, we have $V_A(\phi) = V_A(\psi)$. Our goal is the following. Given a formula ϕ we want to build an *easily readable* formula ψ such that ϕ and ψ are equivalent. We need to explain what we mean by "easily readable". The idea is that the original formula ϕ can be quite complicated and hence we would like to rewrite it in a simpler form, which we refer to as a normal form. Below we formalize this idea.

We start with the notion of literal. A **literal** is a proposition or a negation of a proposition. Thus, every literal is of the form p or $\neg p$, where $p \in PROP$.

Definition 24.1. We say that a formula ψ is in **disjunctive normal form**, or DNF for short, if it is a disjunction of conjunctions of literals. Thus, a formula in disjunctive normal form is written as

$$(\psi_1 \vee \psi_2 \vee \ldots \vee \psi_k),$$

where each ψ_i is a conjunction of literals. Each ψ_i is called a **disjunct** of the formula ψ.

An example of a formula in disjunctive normal form is the following:

$$((p \wedge \neg q) \vee (p \wedge s \wedge \neg t) \vee \neg q \vee r).$$

The formula has 4 disjuncts, which are $(p \wedge \neg q)$, $(p \wedge s \wedge \neg t)$, $\neg q$, and r. The next theorem tells us that every formula can be rewritten in disjunctive normal form.

Theorem 24.1. *Every formula ϕ is equivalent to a formula ψ in a disjunctive normal form.*

Proof.
Let q_1, ..., q_n be all the propositions that occur in ϕ. We write down the truth table for ϕ as in the table earlier. If ϕ is evaluated to **false** at every row then ϕ is equivalent to $(p \wedge \neg p)$, which is clearly in disjunctive normal form. If ϕ is evaluated to **true** at every row then ϕ is equivalent to $(p \vee \neg p)$, which is again in disjunctive normal form. Therefore, we can assume that there are rows (assignments) at which ϕ is evaluated to **true**

and rows at which ϕ is evaluated to **false**. Now we need a small notation. For a proposition q, we write q^1 to mean q and we write q^0 to mean $\neg q$. For the proof we also write 0 for **false** and write 1 for **true**.

Take a row in the truth table. Say the row is the kth row in the truth table, at which ϕ is evaluated to **true**. Let the values of propositions $q_1, \ldots, q_n$ at this row be $a_1, \ldots, a_n$, respectively. Denote this assignment by A_k. Thus each a_i is either **true** or **false**, and $A_k(q_1) = a_1, \ldots, A_k(q_n) = a_n$. Consider the following formula ψ_k, which corresponds to the kth row:

$$(q_1^{a_1} \wedge \ldots \wedge q_n^{a_n}).$$

Thus, for example if all values a_i are **true** then ψ_k is $(q_1 \wedge \ldots \wedge q_n)$, and if all values a_i are **false** then ψ_k is $(\neg q_1 \wedge \ldots \wedge \neg q_n)$. The formula ψ_k has the following three properties:

(1) The formula ψ_k is a conjunction of literals.
(2) The formula ψ_k is evaluated to **true** under A_k.
(3) The formula ψ_k is evaluated to **false** under *all other* truth assignments A.

Let k_1, k_2, ..., k_t be *all* the rows at which ϕ is **true**. Consider the following formula ψ:

$$(\psi_{k_1} \vee \psi_{k_2} \vee \ldots \vee \psi_{k_t}).$$

Clearly, ψ is in disjunctive normal form. We claim that ψ is equivalent to ϕ. In our proof of this claim, we use the three properties above of the formulas ψ_k.

Assume that an assignment A makes ϕ **true**. Then A must appear in a row k in the truth table of ϕ. In that row, ϕ is evaluated to **true**. Thus A is the same as A_k. Therefore, the value of ψ_k is **true** by item (2) above. Since ψ_k appears in ψ, the formula ψ must be **true** under the assignment A.

Assume that an assignment A makes ϕ **false**. Then all the formulas ψ_{k_1}, ψ_{k_2}, ..., ψ_{k_t} are evaluated to **false** under A by item (3) above. Therefore, ψ is evaluated to **false**. Thus, ϕ and ψ are equivalent. □

As an example, consider the formula

$$\phi = ((p \vee \neg q) \rightarrow (s \wedge q)).$$

The following is the truth table for this formula.

p	q	s	ϕ
T	T	T	T
T	T	F	F
T	F	T	F
T	F	F	F
F	T	T	T
F	T	F	T
F	F	T	F
F	F	F	F

Applying the proof of the theorem above, we obtain the following formula in disjunctive normal form equivalent to ϕ:

$$\psi = ((p \wedge q \wedge s) \vee (\neg p \wedge q \wedge s) \vee (\neg p \wedge q \wedge \neg s)).$$

Note that $(p \wedge q \wedge s)$ corresponds to the first row, $(\neg p \wedge q \wedge s)$ corresponds to the fifth row, and $(\neg p \wedge q \wedge \neg s)$ corresponds to the sixth row of the table. These formulas are ψ_1, ψ_5 and ψ_6, respectively, as explained in the proof above.

24.2   The SAT problem

In this section, we discuss one of the most famous problems in mathematics and computer science, **the satisfiability problem**. Roughly the problems asks if a given formula ϕ can be evaluated to **true** under some assignment, and if this can be detected efficiently. We start with the following definition:

Definition 24.2. Let ϕ be a formula.

(1) We say that ϕ is **satisfiable** if there exists a truth assignment that evaluates ϕ to **true**.
(2) We say that ϕ is **valid** if *all* truth assignments evaluate ϕ to **true**. Valid formulas are also called **tautologies**.
(3) We say that ϕ is a **contradiction** if ϕ is not satisfiable.

For example, $((p \vee \neg q) \rightarrow (s \wedge q))$ is satisfiable, but not a tautology. The formula $(\phi \vee \neg \phi)$ is a tautology. For every formula ϕ, it is not too difficult to prove the following proposition.

Proposition 24.1. *Let ϕ be a formula. Then:*

(1) ϕ is a contradiction if and only if $\neg\phi$ is tautology.

(2) ϕ is satisfiable if and only if ϕ is not a contradiction. $\square$

The *SAT* **problem** is formulated as follows. Design an *efficient* algorithm that given a formula ϕ decides whether ϕ is satisfiable. Below we present a simple but inefficient algorithm that solves the problem. The algorithm uses the method $Evaluate(\phi, A)$ that, given a formula ϕ and an assignment A, evaluates ϕ to **true** or **false**.

We say that a sub-formula ψ of ϕ is **basic** if it is in one of the of the following forms

$$\neg p, \ (p \vee q), \ (p \wedge q), \ (p \rightarrow q),$$

where p, q are propositions. Here is an algorithm for the $Evaluate(\phi, A)$ method with A and ϕ being input parameters:

Algorithm 24.1 $Evaluate(\phi, A)$ algorithm

---

(1) Reading ϕ from left to right, find the first occurrence of a basic sub-formula, call it ψ.

(2) Let a represent the value of ψ under the assignment A.

(3) If $\phi = \psi$ then output a and stop.

(4) Set ϕ to be the new formula obtained by replacing ψ with a new proposition q.

(5) Set A to be the old assignment except with the following change: $A(q) = a$.

(6) Repeat the process with the new ϕ and the new A.

In this method, the size of the input formula ϕ is reduced after every iteration. Therefore, the algorithm must terminate. The method is correct as one can prove it by finding a correct loop invariant and induction on the size of the formula (Exercise 4). Consider the next algorithm:

Algorithm 24.2 $SatCheck(\phi)$ algorithm

---

(1) Let $q_1, \ldots, q_n$ be all the propositions in ϕ.

(2) For each assignment A of $q_1, \ldots, q_n$ do

    (a) If $Evaluate(\phi, A)$ outputs **true** then print ϕ *is satisfiable*, and stop.

    (b) If $Evaluate(\phi, A)$ outputs **false** then continue.

(3) Print ϕ *is not satisfiable*.

The algorithm determines if a given formula ϕ is satisfiable. In order to determine this, we list all the truth assignments A of the proposition ϕ, and run $Evaluate(\phi, A)$ for each A. If ϕ contains n propositions then there are 2^n truth assignments. Thus, in the worst case, we run $Evaluate(\phi, A)$ method 2^n times to determine whether ϕ is satisfiable.

Clearly, this algorithm is correct. However, it is a *brute-force* algorithm because it runs through *all* 2^n assignments of the propositions $q_1, \ldots, q_n$. It is a major open problem whether there exists a feasible (polynomial time) solution to the SAT problem. An interesting point of note is this. There is a close relationship between the Hamiltonian curcuit problem discussed in Lecture 8 and the SAT problem. Indeed, if we can find a feasible algorithm to solve the SAT problem then we can efficiently transform that algorithm into a feasible solution for the Hamiltonian curcuit problem. The opposite is also true. Namely, if we can find a feasible algorithm to solve the Hamiltonian curcuit problem then we can efficiently transform that algorithm into a feasible solution for the SAT problem. This is a fascinating topic to studying transformations between the algorithms. However, the topic is beyond the scope of these lectures.

24.3   Models

Consider the set $PROP$ of all propositions. We can think of every subset X of $PROP$ as a truth assignment A_X defined as follows.

$$A_X(p) = \begin{cases} \textbf{true} & \text{if } p \in X; \\ \textbf{false} & \text{otherwise.} \end{cases}$$

In other words, we can view each subset X of $PROP$ as a *model* in which the propositions of X are viewed as **true** and all other propositions are viewed as **false**.

Every truth assignment $A : PROP \rightarrow \{\textbf{true}, \textbf{false}\}$ can be identified with a subset X of $PROP$ in the following way:

$$X = \{p \mid A(p) = \textbf{true}\}.$$

This subset depends on the assignment A, and is therefore denoted by X_A. These observations clearly establish a mapping between subsets of $PROP$ and truth assignments. This mapping sends each subset X of propositions to the truth assignment A_X. This mapping is, in fact, a bijection. In particular, if two sets X and Y of propositions are not equal, then assignments A_X and A_Y are also not equal (as functions).

Let ϕ be a formula. Using the terminology above, we want to define models of ϕ. The models of ϕ will be subsets of $PROP$ defined below, inductively on the length of the formula:

Definition 24.3. Define the **models** of a formula ϕ, denoted by $models(\phi)$, as follows:

Base case: If ϕ is a proposition p, then

$$models(\phi) = \{X \subseteq PROP \mid p \in X\}.$$

Inductive step:

(1) If ϕ is $\neg\phi_1$ then

$$models(\phi) = \{X \subseteq PROP \mid X \notin models(\phi_1)\}.$$

(2) If ϕ is $(\phi_1 \vee \phi_2)$ then

$$models(\phi) = models(\phi_1) \cup models(\phi_2).$$

(3) If ϕ is $(\phi_1 \wedge \phi_2)$ then

$$models(\phi) = models(\phi_1) \cap models(\phi_2).$$

(4) If ϕ is $(\phi_1 \rightarrow \phi_2)$ then

$$models(\phi) = models(\neg\phi_1) \cup models(\phi_2).$$

We say that a subset $X \subseteq PROP$ is a **model** of formula ϕ if $X \in models(\phi)$.

For example, the models of formula p include $\{p\}$, $\{p, q\}$, $\{p, s, q\}$, $\{p, s, \neg r\}$, $\{q, s, t, p\}$, and $\{p, t, s, r\}$. Among these six models $\{p\}$, $\{p, s, \neg r\}$, $\{p, t, s, r\}$ are models of $(p \wedge \neg q)$. The following theorem associates models and truth assignments.

Theorem 24.2. *A subset X is a model of formula ϕ if and only if the truth assignment A_X evaluates ϕ to* **true.**

Proof.
Our proof is by induction on the length of the formula ϕ. In the base case,

when ϕ is a proposition p, it is obvious that X is a model of p if and only if $A_X(p) = $ **true**.

Assume that ϕ is $\neg\phi_1$. Then X is a model of $\neg\phi_1$ if and only if X is not a model of ϕ_1. By the induction hypothesis, X is not a model of ϕ_1 if and only if A_X evaluates ϕ_1 to **false**. Hence, X is a model of ϕ if and only if A_X evaluates $\neg\phi_1$ to **true**.

Assume that ϕ is $(\phi_1 \vee \phi_2)$. Then X is a model of ϕ if and only if X is a model of ϕ_1 or ϕ_2. By the induction hypothesis, X is a model of ϕ if and only if A_X evaluates either ϕ_1 to **true** or ϕ_2 to **true**. Hence X is a model of ϕ if and only if A_X evaluates ϕ to **true**. The other two cases when $\phi = (\phi_1 \wedge \phi_2)$ or $\phi = (\phi_1 \to \phi_2)$ are treated in a similar way. $\square$

24.4　Exercises

(1) How many truth tables are there for formulas whose propositions are $p_1, \ldots, p_n$?

(2) Convert each of the following formulas into disjunctive normal form:

    (a) $(p \to \neg q)$,

    (b) $\neg(p \to (s \to \neg(q \vee p)))$,

    (c) $((\neg p \wedge (s \to p)) \wedge (q \vee \neg s))$,

    (d) $\neg(((p \vee \neg s) \wedge (q \vee \neg p)) \to (p \to q))$.

(3) We say that ψ is in **conjunctive normal form** if it is written as $(\phi_1 \wedge \phi_2 \wedge \ldots \wedge \phi_k)$, where each ϕ is a disjunction of literals. Prove that for every formula ϕ, there exists an equivalent formula ψ in a conjunctive normal form.

(4) Prove Proposition 24.1.

(5) Consider the $Evaluate(\phi, A)$ algorithm. Find a loop invariant of the algorithm and prove that the algorithm is correct.

(6) Finish the inductive step in the proof of Theorem 24.2.

(7) Find all models of the following formulas among all the subsets of $\{p.q, s\}$: $(\neg p \to \neg q)$, $((\neg p \wedge q) \vee (s \to p))$, and $(\neg p \wedge \neg q \wedge \neg s)$.

(8) Prove each of the following:

    (a) If both ϕ and $(\phi \to \psi)$ are valid then ψ is valid.

    (b) If ϕ is valid then so is $(\phi \vee \psi)$.

    (c) If $(\phi \& \psi)$ is valid then both ϕ and ψ are valid.

Programming exercises

(1) Implement the $Evaluate(\phi, A)$ algorithm.
(2) Implement the $SatCheck(\phi)$ algorithm.

Lecture 25

Deterministic finite automata

If you want to be happy, be.
Lev Tolstoy.

25.1 Strings and languages

In this lecture, we introduce deterministic finite automata, one of the simplest mathematical model of computers. Informally, a finite automaton is a system that consists of states and transitions. Each state represents some finite amount of information gathered from the start of the system to the present moment. Transitions represent state changes described by the system rules. Applications of finite automata include digital circuits, sensor data processing, language design, image processing, as well as modeling and building reliable software.

We recall some of the definitions from the first lecture. An alphabet is a finite set Σ of symbols. In most cases, our alphabets contain symbols $a, b, c, d, \ldots$. If Σ contains k letters, then we say that Σ is a k-letter alphabet. A 1-letter alphabet is a **unary alphabet**, and a 2-letter alphabet is a **binary alphabet**. We will often consider binary alphabets.

The elements of the alphabet are called **input symbols**. A finite sequence of symbols from Σ is called a **string** or **word** over the alphabet. Thus, each string is of the form $\sigma_1 \sigma_2 \ldots \sigma_n$, where each σ_i is a symbol of the alphabet. The **length** of the string $\sigma_1 \sigma_2 \ldots \sigma_n$ is n. Thus, the lengths of $aaab$, $bababa$, and bb are 4, 6, and 2, respectively. There is a special string, whose length is 0, called the **empty string**. We denote this string by λ. Using mathematical induction, it is easy to show that there are k^n strings of length n over a k-letter alphabet.

The set of all strings over the alphabet Σ is denoted by $\Sigma^\star$:

$$\Sigma^\star = \{\sigma_1\sigma_2\ldots\sigma_m \mid \sigma_1, \sigma_2, \ldots, \sigma_m \in \Sigma, \ m \in \mathbb{N}\}.$$

Note that when $m = 0$, we have the empty string λ and therefore $\lambda \in \Sigma^\star$. We denote strings by the letters $u, v, w, \ldots$.

The **concatenation** of two strings u and v, denoted by $u \cdot v$, is obtained by writing u followed by v. For example, $aab \cdot bba$ produces $aabbba$. The concatenation operation on words satisfies the equality, known as the *associativity law*,

$$u \cdot (v \cdot w) = (u \cdot v) \cdot w$$

for all words u, v, w. Note that $\lambda \cdot u = u \cdot \lambda$ for any string u. Often, instead of writing $u \cdot v$, we omit the dot $\cdot$ sign and write uv.

Let u be a string. The notation u^n represents the string obtained by writing u exactly n times. Thus,

$$u^n = \underbrace{u \cdot u \cdot \ldots \cdot u}_{n \ \ times}.$$

For example, $(ab)^3 = ababab$. When $n = 0$, then u^n is the empty string λ. We say that w is a **substring** of u if w occurs in u. Formally, w is a substring of u if $u = u_1 w u_2$ for some strings u_1 and u_2. For example, ab is a substring of $aaabbbbaa$, whereas aba is not. Clearly, every string u is a substring of itself. We say that w is a **prefix** of u if u can be written as $w u_1$, for some string u_1. For example, the prefixes of the string $abbab$ are $\lambda, a, ab, abb, abba$, and $abbab$.

Definition 25.1. A **language** (over alphabet Σ) is a subset of $\Sigma^\star$.

When we use the term language we always assume, implicitly or explicitly, that we are given an alphabet. Here are some examples of languages:

$$\emptyset, \ \Sigma^\star, \ \{a^n \mid n \in \mathbb{N}\}, \ \{ab, ba\}, \ \{w \in \{a, b\}^\star \mid aba \text{ is a substring of } w\}, \ \{\lambda\}.$$

25.2 Operations on languages

We denote languages by capital letters U, V, W, L, etc. We now define several operations on languages. Most of the operations recast the set-theoretic operations of union, intersection, and complementation explained in Lecture 10 on sets.

Let U and V be languages of the alphabet Σ. The following are the *Boolean operations* on languages:

(1) The **union** of U and V is $U \cup V$,
(2) The **intersection** of U and V is is $U \cap V$; and
(3) The **complement** of U is $\Sigma^* \setminus U$.

Below we discuss two additional operations specific to languages.

The concatenation operation. Let U and W be languages. The concatenation of U and W, denoted by $U \cdot V$, is the language

$$U \cdot V = \{u \cdot w \mid u \in U, w \in W\}.$$

For example, if $U = \{ab, ba\}$ and $W = \{aa, bb\}$ then $U \cdot W = \{abaa, abbb, baaa, babb\}$. For $U = \{a\}$ and any language L, we have $U \cdot L = \{au \mid u \in L\}$.

The star operation. This operation is often also called the Kleene's star operator. Let U be a language. Consider the following sequence of languages: $U^0, U^1, U^2, U^3, U^4, \ldots$, where the language U^n with $n \in \mathbb{N}$ is defined recursively by the rule:

$$U^0 = \{\lambda\}, \ U^1 = U, \ U^2 = U \cdot U, \text{ and } U^n = U^{n-1} \cdot U.$$

Take the union of all these languages and denote the resulting language by U^*. Thus,

$$U^* = U^0 \cup U^1 \cup U^2 \cup U^3 \cup \ldots.$$

The language U^* is called **the star of the language** U. In other words, U^* is the set of strings obtained by finite concatenations applied to strings from U. For example, if $U = \{a\}$ then $U^* = \{\lambda, a, aa, aaa, \ldots\}$. The star of every language contains λ. The language U^* is always infinite apart from the cases $U = \emptyset$ and $U = \{\lambda\}$.

25.3 Deterministic finite automata (DFA)

Let Σ be an alphabet and U be a language of Σ. A typical problem that we want to solve is the following. Design an algorithm that, given a string w, determines whether or not w is in U.

Here is an example. Consider the language

$$U = \{w \in \{a, b\}^* \mid w \text{ contains the substring } aba\}.$$

We want to design an algorithm that, given a string w, determines whether $w \in U$. If U were a finite language, then writing such a program would be easy. The program would consist of a set of *if* statements, one for each string in U, checking whether w were equal to that string or not. However, U is an infinite set, and thus writing a finite program is non-trivial. Here is the *Find-aba(w)*-algorithm which performs this check. Let w be an input string $w = \sigma_1 \ldots \sigma_n$.

(1) Initialize variables i and *state* as $i = 1$ and *state* $= 0$.
(2) If *state* $= 0$ and $\sigma_i = b$ then set *state* $= 0$.
(3) If *state* $= 0$ and $\sigma_i = a$ then set *state* $= 1$.
(4) If *state* $= 1$ and $\sigma_i = a$ then set *state* $= 1$.
(5) If *state* $= 1$ and $\sigma_i = b$ then set *state* $= 2$.
(6) If *state* $= 2$ and $\sigma_i = a$ then set *state* $= 3$.
(7) If *state* $= 2$ and $\sigma_i = b$ then set *state* $= 0$.
(8) If *state* $= 3$ and $\sigma_i \in \{a, b\}$ then *state* $= 3$.
(9) Increment i by one.
(10) If $i = n + 1$ then go to Line 11. Otherwise go to Line 2.
(11) If *state* $= 3$ then output *accept*. Otherwise output *reject*.

The important feature of this algorithm is the set of values assigned to the variable *state*. These values are 0, 1, 2, and 3. State 0 represents the initial state. State 1 indicates that symbol a has just been seen. State 2 represents that the string ab has just been read. Finally, state 3 signals that aba has been detected. We call these the states of the *Find-aba(w)*-algorithm. The algorithm makes its transitions from one state to another depending on the input σ read. Thus, we have a transition function T associated with the program presented in the table below.

T	a	b
0	1	0
1	1	2
2	3	0
3	3	3

The state 0 is the *initial state* of the algorithm. The state 3 is the *accepting state* as w contains aba when *state* $= 3$. Finally, if the algorithm outputs *reject* then the *state* variable has value 0, 1, or 2. Thus, we have given a finite state analysis of the *Find-aba(w)* algorithm. Now, we can abstract from this example and give the following definition:

Definition 25.2. **A deterministic finite automaton (DFA)** is a 5-tuple (S, q_0, T, F, Σ), where

(1) S is the set of **states**,
(2) q_0 is the **initial state** and $q_0 \in S$,
(3) T is the **transition function** $T : S \times \Sigma \to S$,
(4) F is a subset of S called the set of **accepting states**, and
(5) Σ is an alphabet.

We often write (S, q_0, T, F) instead of (S, q_0, T, F, Σ) since Σ is given by the transition function T. We use letters $\mathcal{A}$, $\mathcal{B}$, ... to denote finite automata.

Thus, the state analysis of the $Find\text{-}aba(w)$ algorithm above gives us the automaton $(S, 0, T, F)$, where $S = \{0, 1, 2, 3\}$, 0 is the initial state, T is presented in the table above, and $F = \{3\}$.

We can visualize finite automata as labeled directed graphs. Let (S, q_0, T, F) be a DFA. The states of the DFA are represented as vertices of the graph. We put an edge from state s to state q if there is an input signal σ such that $T(s, \sigma) = q$, and we label this edge with σ. The edges labeled with σ are called σ-**transitions**. The initial state is presented as a vertex with an ingoing arrow without a source. The accepting states are vertices that are filled in grey. Sometimes the accepting states are denoted by drawing a double ring. We call this visual presentation a **transition diagram** of the automaton. For example, the transition diagram of the automaton for the $Find\text{-}aba(w)$ algorithm is given in Figure 25.1. Another example of a DFA is presented in Figure 25.2.

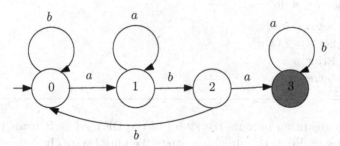

Fig. 25.1: Transition diagram for a DFA for the $Find\text{-}aba(w)$ algorithm

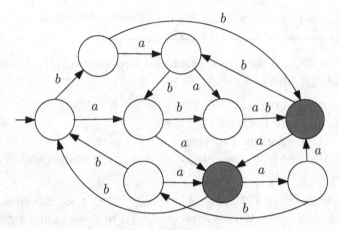

Fig. 25.2: An example of a transition diagram

Definition 25.3. Let $\mathcal{M} = (S, q_0, T, F, \Sigma)$ be a DFA and $w = \sigma_1 \ldots \sigma_n$ be a string. The **run** of the automaton on this string is the sequence of states $s_1, s_2, \ldots, s_n, s_{n+1}$ such that s_1 is the initial state and $T(s_i, \sigma_i) = s_{i+1}$ for all $i = 1, \ldots, n$.

The run of $\mathcal{M}$ on string $w = \sigma_1 \ldots \sigma_n$ can be viewed as the execution of the following algorithm $Run(\mathcal{M}, w)$:

Algorithm 25.1 $Run(\mathcal{M}, w)$ algorithm

(1) Initialize $s = q_0$, $i = 1$, and print s.
(2) *While* $i \leq n$ do

    (a) Set $\sigma = \sigma_i$.
    (b) Set $s = T(s, \sigma)$.
    (c) Print s.
    (d) Increment i.

This algorithm outputs the states of the DFA $\mathcal{M}$ as it reads through the string w. First, the algorithm prints the initial state. It then reads the first input symbol σ_1, transitions from the initial state to the state $T(q_0, \sigma_1)$ and outputs $T(q_0, \sigma_1)$, reads the next symbol σ_2, transitions again, and so on. Every run of $\mathcal{M}$ is a path starting from the initial state q_0. In the transition diagram of $\mathcal{M}$, one can visualize the run as a path labeled by the string w and starting with q_0. For example, the path

$$0, 0, 0, 1, 1, 2, 3, 3$$

is the run of the automaton in Figure 25.1 on the string *bbaabab*.

Definition 25.4. We say that $\mathcal{M}$ **accepts** the string $w = \sigma_1 \ldots \sigma_n$ if the run $s_1, \ldots, s_n, s_{n+1}$ of M on w is such that the last state $s_{n+1} \in F$. We call such a run an **accepting run**.

Note that the run of $\mathcal{M}$ on w is always unique. This can be proved on induction on the length of the input string. See Exercise 5. Hence the run is either accepting or rejecting but not both. For example, the automaton in Figure 1 accepts those strings that contain *aba* as a substring and rejects all others. We define the following central concept:

Definition 25.5. Let $\mathcal{M} = (S, q_0, T, F, \Sigma)$ be a DFA. The **language accepted by** $\mathcal{M}$, denoted by $L(\mathcal{M})$, is the following language:

$$L(\mathcal{M}) = \{w \mid \text{the automaton } \mathcal{M} \text{ accepts } w\}.$$

A language $L \subseteq \Sigma^\star$ is **DFA recognizable** if there exists a DFA $\mathcal{M}$ such that $L = L(\mathcal{M})$.

We give several simple examples:

(1) Consider a DFA with exactly one state. If the state is the accepting state then the automaton accepts the language $\Sigma^\star$. Otherwise, the automaton recognizes the empty language $\emptyset$.

(2) Consider the language $\{w\}$ consisting of one word $w = \sigma_1 \ldots \sigma_n$. This language is recognized by the following DFA $(S, 0, T, F)$:

   (a) $S = \{0, 1, 2, 3, 4, \ldots, n+1\}$ with 0 being the initial state.
   (b) For all $i \leq n-1$, $T(i, \sigma_{i+1}) = i+1$. In all other cases $T(s, \sigma) = n+1$.
   (c) The accepting state is n.

   The transition diagram of this automaton when $w = abbab$ is depicted in Figure 25.3.

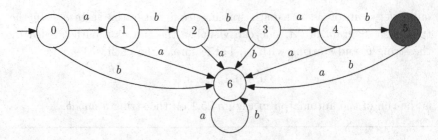

Fig. 25.3: A DFA recognizing the language $\{abbab\}$.

(3) The language $L = \{aw \mid w \in \{a,b\}^\star\}$ is DFA recognizable. The transition diagram of a DFA (S, q_0, T, F) that recognizes the language is in Figure 25.4. Formally:

(a) $S = \{0, 1, 2\}$ with 0 being the initial state.
(b) The transition table is defined as follows: $T(s, \sigma) = 1$ for $s = 0$ and $\sigma = a$, $T(s, \sigma) = 1$ for $s = 1$ and $\sigma \in \{a, b\}$, and $T(s, \sigma) = 2$ in all other cases.
(c) 1 is the accepting state.

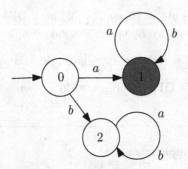

Fig. 25.4: A DFA recognizing the language $\{aw \mid w \in \{a,b\}^\star\}$.

25.4   Exercises

(1) Let Σ be a k-letter alphabet. Prove that for any given n, the number of strings of length n over this alphabet is equal to k^n.
(2) On the set $\Sigma^\star$ consider the following relations:

(a) *SubString* $= \{(w, u) \mid w$ is a substring of $u\}$.

(b) $Prefix = \{(w, u) \mid w \text{ is a prefix of } u\}$

Prove that the *SubString* and *Prefix* relations determine partial orders on the set Σ^*.

(3) Prove each of the following:

    (a) The concatenation operation on the set of all strings satisfies the associativity law.

    (b) The concatenation operation on the set of all languages satisfies the associativity law.

(4) Prove that $U^* \cup V^* \subseteq (U \cup V)^*$. Find several conditions on U and V that guarantee the equality $U^* \cup V^* = (U \cup V)^*$.

(5) Let $\mathcal{M}$ be a DFA. Prove that the run of $\mathcal{M}$ on any given string w is unique.

(6) Prove that every finite language is recognized by a deterministic finite automaton.

(7) Draw transition diagrams of DFAs that recognize the following languages. The alphabet Σ in this example is $\{a, b\}$.

    (a) $\{wb \mid w \in \Sigma^*\}$.

    (b) $\{\lambda\}$.

    (c) $\{w \mid w \in \Sigma^* \text{ and } w \neq \lambda\}$.

    (d) $\{uaabv \mid u, v \in \Sigma^*\}$.

    (e) $\{a, b\}^*$.

    (f) $\{a\}^* \cdot \{b\} \cdot b \cdot \Sigma^*$.

    (g) $\{w \mid \text{the length of } w \text{ is a multiple of } k\}$, where k is fixed.

(8) Draw a transition diagram of a DFA that recognizes the following language: $L = \{w \mid w \text{ is a binary string that has 0 at the third position from the right of } w\}$.

Programming exercises

(1) Design a class that represents DFA. Make sure that

    (a) the class has a constructor that takes as input a set of states, a start state, a transition table, and a set of final states.

    (b) the class contains a method which determines if the DFA accepts or rejects a given string.

Lecture 26

Designing finite automata

Everything is designed. Few things are designed well.

Brian Reed.

26.1 Two examples

Suppose that we are given a language L and we are asked to design a DFA that recognizes the language L or argue that such a DFA does not exist. How do we solve this problem?

Example 1. Suppose we want to construct a deterministic finite automaton that recognizes the language

$$L = \{w \mid w \in \{a, b\}^\star \text{ and } w \text{ contains an odd number of } a\text{'s and an even number of } b\text{'s}\}.$$

We start reading an input string w from left to right. Do we need to remember the *entire* string in order to tell whether we have passed through an odd number of a's and an even number of b's? No, because of the following observation. Suppose we use two coins to keep track of what we have seen so far. We associate the first coin with the symbol a and the second with b. When we start, both coins are on *heads*. Reading w, every time we see a we flip the first coin, and every time we see b we flip the second coin. Thus, if the first coin is on *heads* then we have seen an even number of a's, and if the coin is on *tails* then we have seen an odd number of a's. The same holds true for the symbol b and the second coin. Therefore, at any given time, our state is determined by the current sides of the coins. There are four possible states:

(1) Both coins are on *heads*. Denote this state by 0.

(2) The first coin is on *tails* and the other is on *heads*. Denote this state by 1.

(3) The first coin is on *heads* and the other is on *tails*. Denote this state by 2.

(4) Both coins are on *tails*. Denote this state by 3.

When we finish reading the string w, we accept the string if the first coin is on *tails* and the second is on *heads*. In all other cases, we reject the string. This reasoning helps us to build a DFA that recognizes L. Figure 26.1 depicts the transition diagram of the automaton.

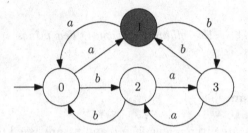

Fig. 26.1: A DFA recognizing the language L in Example 1.

Example 2. Consider the language

$$L = \{a^n b^n \mid n \in N\}.$$

We would like to reason that no DFA recognizes L. For now, we keep the discussion informal. We revisit this example in Lecture 30 that makes the reasoning below formal. While processing an input string w, we must read w from left to right without going back to any symbol that we have already read. Assume w starts with b. Then, we reject w. Now assume that w starts with an a. We read a and remember that we have read one a. This tells us that we have to create a state, let us denote it by s_1, that remembers that w starts with an a. If b is the next and the last symbol then we accept w. If the second symbol is a, then we remember that we have seen two a's. This tells us that we have to create a new state s_2 that remembers that w starts with aa. We must keep s_1 and s_2 separate. Otherwise, we would, incorrectly, accept the string aab which is not in the language. We continue this process further. Assume that w is of the form $a^n u$, where $n \geq 1$, and we have created states $s_1, s_2, \ldots, s_n$. State s_i detects that the input string starts with a^i. Consider the first symbol of u. If it is a b then we can traverse our states $s_n, s_{n-1}, \ldots, s_1$ as we read the next n symbols of w. If all the symbols of u are b and there are exactly n of them, we accept w and otherwise reject. However, if the first symbol of u is a, then we need to remember that we have read exactly $n + 1$ symbols of a. This tells us that we have to create a new state s_{n+1} that remembers that w starts with a^{n+1}. One can argue that we must keep s_{n+1} separate from all the states s_1, s_2, ..., s_n we have built so far. Thus, we see that in order to recognize L, our informal analysis tells us that we need to create infinitely many states.

26.2 Constructing automata for the complementation operation

Assume that we are given a DFA $\mathcal{M} = (S, q_0, T, F)$. Our goal is to design a DFA that recognizes the complement of $L(\mathcal{M})$, that is, we want to construct a DFA $\mathcal{M}^{(c)}$ such that for all strings w the DFA $\mathcal{M}$ does not accept w if and only if the DFA $\mathcal{M}^{(c)}$ accepts w. This turns out to be a simple procedure. To construct the desired DFA $\mathcal{M}^{(c)}$, we use the original DFA $\mathcal{M}$. Namely, we keep the set of states the same, we also keep the initial state, and the transition function T. The only change we make is that we swap the accepting states with the non-accepting states. Thus, the automaton that recognizes the language $\Sigma^{\star} \setminus L(\mathcal{M})$ is $\mathcal{M}^{(c)} = (S, q_0, T, S \setminus F)$. Here the following observation is important. On *every* input w, the DFA $\mathcal{M}$ (and hence $\mathcal{M}^{(c)}$) has a *unique* run. Therefore, $\mathcal{M}$ accepts w if and only if $\mathcal{M}^{(c)}$ rejects w.

26.3 Constructing automata for the union operation

Assume we are given two deterministic finite automata

$$\mathcal{M}_1 = (S_1, q_0^{(1)}, T_1, F_1) \text{ and } \mathcal{M}_2 = (S_2, q_0^{(2)}, T_2, F_2).$$

These two DFA recognize languages $L_1 = L(\mathcal{M}_1)$ and $L_2 = L(\mathcal{M}_2)$. We want to design a DFA $\mathcal{M} = (S, q_0, T, F)$ that recognizes the union $L_1 \cup L_2$.

Our initial idea for building the desired $\mathcal{M}$ is this. For an input string w, we simulate the first machine $\mathcal{M}_1$. If $\mathcal{M}_1$ accepts w then $w \in L_1 \cup L_2$. If $\mathcal{M}_1$ rejects w then we run the second machine $\mathcal{M}_2$ on w. If $\mathcal{M}_2$ accepts w then $w \in L_1 \cup L_2$. If $\mathcal{M}_2$ does not accept w, then $w \notin L_1 \cup L_2$. The problem with this idea is that finite automata cannot read the input w twice. Instead we run $\mathcal{M}_1$ and $\mathcal{M}_2$ on w *in parallel*.

We run $\mathcal{M}_1$ and $\mathcal{M}_2$ simultaneously on input string w as follows. We start at the initial states $q_0^{(1)}$ and $q_0^{(2)}$. We read the first input symbol σ_1 of w, make simultaneous σ_1-transitions on $\mathcal{M}_1$ and $\mathcal{M}_2$, and remember the states $s_1^{(1)} = T_1(q_0^{(1)}, \sigma_1)$ and $s_1^{(2)} = T_2(q_0^{(2)}, \sigma_1)$. We then repeat this process by making simultaneous transitions from $s_1^{(1)}$ to $s_2^{(1)} = T_1(s_1^{(1)}, \sigma_2)$ on $\mathcal{M}_1$, and from $s_1^{(2)}$ to $s_2^{(2)} = T_2(s_1^{(2)}, \sigma_2)$ on $\mathcal{M}_2$, where σ_2 is the second letter of w. We continue this, and once we finish the simultaneous runs of $\mathcal{M}_1$ and $\mathcal{M}_2$ on w, we look at the resulting end states of these two runs. If one of these states is accepting then we accept w, otherwise we reject w. Thus, at any given stage of the two runs we remember a pair (p, q), where $p \in S_1$ and $q \in S_2$. Our transition on this pair on input σ is the simultaneous transitions from p to $T_1(p, \sigma)$ of $\mathcal{M}_1$ and from q to $T_2(q, \sigma)$ of $\mathcal{M}_2$.

Now, we formally define the DFA $\mathcal{M} = (S, q_0, T, F)$ that recognizes the language $L_1 \cup L_2$. Note that in our construction we use the product of transition functions defined in Section 21.4.

(1) The set S of states is $S_1 \times S_2$.
(2) The initial state is the pair $(q_0^{(1)}, q_0^{(2)})$.
(3) The transition function T is the product of the transition functions T_1 and T_2, that is:
$$T((p, q), \sigma) = (T_1(p, \sigma), T_2(q, \sigma)),$$
where $p \in S_1$, $q \in S_2$, and $\sigma \in \Sigma$.
(4) The set F of accepting states consists of all pairs (p, q) such that either $p \in F_1$ *or* $q \in F_2$.

To prove that $\mathcal{M}$ is a DFA recognizing the union $L_1 \cup L_2$, we show that, on the one hand, if w is in $L_1 \cup L_2$ then the automaton $\mathcal{M}$ accepts w. Indeed, if $w \in L_1 \cup L_2$ then either $\mathcal{M}_1$ accepts w or $\mathcal{M}_2$ accepts w. In either case, since $\mathcal{M}$ simulates both $\mathcal{M}_1$ and $\mathcal{M}_2$, the string w must be accepted by $\mathcal{M}$. On the other hand, if w is accepted by $\mathcal{M}$ then the run of $\mathcal{M}$ on w is split into two runs: one is the run of $\mathcal{M}_1$ on w and the other is the run of $\mathcal{M}_2$ on w. Since $\mathcal{M}$ accepts w, it must be the case that one of the runs is accepting.

The notation for the automaton built is $\mathcal{M}_1 \oplus \mathcal{M}_2$.

For instance, consider the automata $\mathcal{M}_1$ and $\mathcal{M}_2$ that recognize the languages L_1 and L_2, respectively, over the alphabet $\{a, b\}$: $L_1 = \{v \mid v$ contains an odd number of a's$\}$ and $L_2 = \{v \mid v$ contains an even number of b's$\}$. The automaton $\mathcal{M}_1$ recognizing L_1 is defined as follows:

(1) $S_1 = \{s_0, s_1\}$,

(2) s_0 is the initial state and s_1 is the accepting state.

(3) $T_1(s_0, b) = s_0$, $T(s_0, a) = s_1$, $T_1(s_1, b) = s_1$, and $T(s_1, a) = s_0$.

The automaton $\mathcal{M}_2$ recognizing L_2 is defined as follows:

(1) $S_2 = \{q_0, q_1\}$,

(2) q_0 is the initial state and the accepting state.

(3) $T_2(q_0, b) = q_1$, $T_2(q_0, a) = q_0$, $T_2(q_1, b) = q_0$, and $T_2(q_1, a) = q_1$.

The automaton $\mathcal{M}_1 \oplus \mathcal{M}_2$ that recognize the language $L_1 \cup L_2$ is then the following:

(1) The set of states is $S_1 \times S_2 = \{(s_0, q_0), (s_0, q_1), (s_1, q_0), (s_1, q_1)\}$.

(2) The initial state is (s_0, q_0).

(3) The accepting states are (s_0, q_0), (s_1, q_0), and (s_1, q_1).

(4) The transition table T is determined as follows:

$$T((p, q), \sigma) = (T_1(p, \sigma), T_2(q, \sigma)),$$

where $p \in S_1$, $q \in S_2$, and $\sigma \in \Sigma$. For instance $T((s_1, q_0), a) = (s_0, q_0)$ and $T((s_0, q_0), b) = (s_0, q_1)$.

26.4 Constructing automata for the intersection operation

Assume we are given two DFA $\mathcal{M}_1 = (S_1, q_0^{(1)}, T_1, F_1)$ and $\mathcal{M}_2 = (S_2, q_0^{(2)}, T_2, F_2)$. These two DFA recognize languages $L_1 = L(\mathcal{M}_1)$ and $L_2 = L(\mathcal{M}_2)$. We want to design a DFA $\mathcal{M} = (S, q_0, T, F)$ that recognizes the intersection $L_1 \cap L_2$. We modify the idea of constructing the DFA for the union of languages. Given an input w, we run $\mathcal{M}_1$ and $\mathcal{M}_2$ on w simultaneously as we explained for the union operation. Once we finish the runs of $\mathcal{M}_1$ and of $\mathcal{M}_2$ on w, we look at the resulting end states of these two runs. If *both* end states are accepting then we accept w, otherwise we reject w. Formally, we define the DFA $\mathcal{M} = (S, q_0, T, F)$ that recognizes the language $L_1 \cap L_2$ as follows:

(1) The set S of states is $S_1 \times S_2$.

(2) The initial state is the pair $(q_0^{(1)}, q_0^{(2)})$.

(3) The transition function T is the product of the transition functions T_1 and T_2, that is:

$$T((p, q), \sigma) = (T_1(p, \sigma), T_2(q, \sigma)),$$

where $p \in S_1$, $q \in S_2$, and $\sigma \in \Sigma$.

(4) The set F of final states consists of all pairs (p, q) such that $p \in F_1$ *and* $q \in F_2$.

The only difference between $\mathcal{M}$ and $\mathcal{M}_1 \oplus \mathcal{M}_2$ is in the accepting states. The notation for the automaton $\mathcal{M}$ is $\mathcal{M}_1 \otimes \mathcal{M}_2$.

It is easy to prove that the constructed DFA $\mathcal{M}$ recognizes $L_1 \cap L_2$. Indeed, assume that a word w is accepted by both $\mathcal{M}_1$ and $\mathcal{M}_2$. Let $s_0, s_1, \ldots, s_n$ be the accepting run of $\mathcal{M}_1$ on w and let $p_0, \ldots, p_n$ be the accepting run of $\mathcal{M}_2$ on w. Then the sequence

$$(s_0, p_0), (s_1, p_1), \ldots, (s_n, p_n)$$

is the run of $\mathcal{M}_1 \otimes \mathcal{M}_2$ on w. Since both s_n and p_n are accepting states of $\mathcal{M}_1$ and $\mathcal{M}_2$, respectively, we have that (s_n, p_n) is an accepting state of $\mathcal{M}_1 \otimes \mathcal{M}_2$. Hence w is accepted by $\mathcal{M}_1 \otimes \mathcal{M}_2$. Now assume that w is accepted by $\mathcal{M}_1 \otimes \mathcal{M}_2$. Let

$$(s_0, p_0), (s_1, p_1), \ldots, (s_n, p_n)$$

be the accepting run of $\mathcal{M}_1 \otimes \mathcal{M}_2$ on w. Then by the definition of T, we have that $s_0, s_1, \ldots, s_n$ and $p_0, \ldots, p_n$ are accepting runs of $\mathcal{M}_1$ and $\mathcal{M}_2$ on w, respectively. Hence w belongs to L_1 and L_2.

For instance, consider the automata $\mathcal{M}_1$ and $\mathcal{M}_2$ that recognize the languages L_1 and L_2, respectively, over the alphabet $\{a, b\}$: $L_1 = \{v \mid v$ contains an odd number of a's$\}$ and $L_2 = \{v \mid v$ contains an even number of b's $\}$.

We have already constructed the automata $\mathcal{M}_1$ and $\mathcal{M}_2$ for these languages in the previous section. The automaton $\mathcal{M}_1 \otimes \mathcal{M}_2$ then recognizes the language $L = \{w \mid w \in \{a, b\}^*$ and w contains an odd number of a's and an even number of b's$\}$. In Section 1 in Example 1, we already built an automaton recognizing this language. Notice that the transition diagram of the automaton $\mathcal{M}_1 \otimes \mathcal{M}_2$ is the same as in Figure 26.1 with states renamed.

26.5   Exercises

(1) Construct the union, intersection and complementation automata for the following languages:

   (a) $\{uabv \mid u, v \in \Sigma^*\}$ and $\{ubbv \mid u, v \in \Sigma^*\}$.
   (b) $\{u \mid$ the length of u equals 0 modulo 3$\}$ and $\{u \mid u$ contains even number of $a\}$.

(2) For every positive integer n give an example of a language L_n such that no DFA with less than n states recognizes L_n. Can L_n be a finite language?

(3) Draw the transition diagrams of DFA recognizing the following languages over the alphabet $\{a, b\}$.

(i) $\{w \mid w \in \{ab\}^\star\}$

(ii) $\{w \mid w$ has r number of a's modulo $p\}$, where r and p are both fixed.

(iii) $\{aw \mid w \in \{a, b\}^\star\}$.

(4) Let $\mathcal{A}$ be a DFA. Consider the DFA $\mathcal{B}$ obtained by removing all the states in $\mathcal{A}$ that are not reachable from the initial state of $\mathcal{A}$. Prove that $\mathcal{A}$ and $\mathcal{B}$ are equivalent, that is $L(\mathcal{A}) = L(\mathcal{B})$.

(5) Consider the unary alphabet $\Sigma = \{a\}$. A language L of this alphabet is called **ultimately periodic** if there are natural numbers n and $p \geq 0$ such that

$$L = \{a^n, a^{n+p}, a^{n+2p}, a^{n+3p}, \ldots\}.$$

Prove the following:

(a) Every ultimately periodic language is FA recognizable.

(b) If $L \subseteq \Sigma^\star$ is FA recognizable then L is a finite union of ultimately periodic and finite languages.

(6) Consider the construction of the DFA $\mathcal{M}_1 \oplus \mathcal{M}_2$ in Section 26.3. Give a more formal proof that $L(\mathcal{M}_1 \oplus \mathcal{M}_2) = L(\mathcal{M}_1) \cup L(\mathcal{M}_2)$.

Programming exercises

(1) Implement the following methods:

(a) A method that takes two DFA $\mathcal{M}_1$ and $\mathcal{M}_2$ as input and outputs $\mathcal{M}_1 \oplus \mathcal{M}_2$.

(b) A method that takes two DFA $\mathcal{M}_1$ and $\mathcal{M}_2$ as input and outputs $\mathcal{M}_1 \otimes \mathcal{M}_2$.

(c) A method that takes a DFA $\mathcal{M}$ as an input and outputs $\mathcal{M}^{(c)}$.

Lecture 27

Nondeterministic finite automata

Sail away from the safe harbor. Catch the trade
winds in your sails. Explore. Dream. Discover.

Mark Twain.

27.1 Definitions and examples

We start with an example. For a fixed natural number $n \geq 1$, consider the language

$$L = \{uav \mid u, v \in \{a, b\}^\star \text{ and the length of } v \text{ is } n - 1\}.$$

Thus, L consists of all strings w that have an a in the nth position from the right of w. Assume that we are asked to design a DFA recognizing L and draw the transition diagram of the automaton. When trying to design a DFA recognizing this language, one eventually realizes that such a DFA exists and that it has many states. We will prove later on that there exists a DFA with 2^n states recognizing this language. Moreover, no DFA with less than 2^n states recognizes L. Below we present an alternative approach to recognize this language. This approach leads us to the definition of nondeterministic finite automata (NFA).

Imagine that we are reading an input string w from $\{a, b\}^\star$ to determine whether w is in L. If we do not see the symbol a then we reject w. Let us assume that we have just seen some a in w. We can make one of the following two *guesses*:

(1) The symbol a we have just seen is in the nth position from the right of w.

(2) The symbol a we have just seen is *not* in the nth position from the right of w.

Suppose we select the first guess. We change our state, and then try to verify our guess. Let us call this state the *verification* state. To verify the correctness of our guess we just need to read the rest of the string and check if the length of the rest of the string is $n - 1$. To do this, we set the integer variable *length* to 0 and increment it by 1 each time we read the next input symbol of w. If the length of the rest of the string is strictly less than $n - 1$, then the guess is incorrect. Now assume that $length = n - 1$. If the string does *not* have any more symbols left then the guess is correct. Otherwise, the guess is incorrect. In this case, we need to go back to the position of w where we made the first guess, and select the second guess.

Suppose we decide to select the second guess. Call this state the *passive* state. This means we continue on with the rest of the string just passing through all the b's until we encounter the next a. When we see the next a, we repeat the same guessing process by selecting either the *verification* state or the *passive* state. Thus, while reading w, whenever we see an a we can either stay in the *passive* state or we can select the *verification* state and verify the correctness of our guess.

An important note is this. If $w \in L$ then w is of the form uav where the length of v is $n - 1$. Therefore, we can act as follows. We stay in the *passive* state until we read through the entire string u. Once we finish reading u, we select the *verification* state and easily verify the correctness of our guess. Now suppose that $w \notin L$. Then *all* our guesses when we select the *verification* state will never confirm that our guess was correct.

The *passive* state is **nondeterministic** in the following sense. When we are in the *passive* state and read an a, it is *not* determined which of the following two transition we should make. One is to stay in the *passive* state, and the other is to move to the *verification* state. In Figure 28.1 we represent our analysis above by a finite state machine for the case when $n = 3$. In this figure the initial state is the *passive state* and the state after the initial state is the *verification state*. Notice that there are two possible transitions out of the initial state 0 on input symbol a.

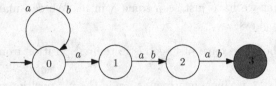

Fig. 27.1: An example of an NFA

Keeping this example in mind, we give the following central definition:

Definition 27.1. A **nondeterministic finite automaton** is a 5-tuple (S, I, T, F, Σ), where

(1) S is the set of **states**,
(2) $I \subseteq S$ is the set of initial states,
(3) T is the **transition function** $T : S \times \Sigma \to P(S)$, where $P(S)$ is the power set of S,
(4) $F \subseteq S$ is the set of **accepting states**, and
(5) Σ is a given alphabet.

We abbreviate nondeterministic finite automata by NFA. We often write (S, I, T, F) instead of (S, I, T, F, Σ) as it will be clear from T which Σ is used. We use letters $\mathcal{A}$, $\mathcal{B}$, ... to denote NFA. We often say **automata** for short, instead of nondeterministic finite automata.

We can visualize a nondeterministic finite automaton (S, I, T, F) by a labeled directed graph, similar to our presentations of DFA. The states of the NFA are vertices of the graph. We put an edge from state s to state p and label it with σ if $p \in T(s, \sigma)$. These are called σ-**transitions**. The initial states are the vertices that have incoming arrows without a source. The accepting states are filled in grey. We call this visual presentation a **transition diagram** of the automaton. Since $T : S \times \Sigma \to P(S)$ it may well be the case that $T(s, \sigma) = \emptyset$ for some $s \in S$ and $\sigma \in \Sigma$.

Every DFA can clearly be viewed as an NFA. The difference between a DFA and NFA is that every state in a DFA is required to always have *exactly* one σ-transition for every $\sigma \in \Sigma$. An NFA, on the other hand, can contain states that have zero, one, two or more σ-transitions. States with more than one σ-transitions are called **nondeterministic** states.

The NFA presented in Figure 27.2 is formally defined as follows:

(1) $S = \{0, 1, 2, 3, 4, 5, 6\}$ and $F = \{2, 5\}$.
(2) $I = \{0, 6\}$.
(3) The table of the transition function T is defined as follows:

- $T(0, a) = \{0, 1\}$, $T(0, b) = \{0\}$, $T(1, a) = T(1, b) = \{1\}$,
- $T(2, a) = \{3\}$, $T(2, b) = \emptyset$, $T(3, a) = T(3, b) = \{4\}$,
- $T(4, a) = T(4, b) = \{5\}$, $T(5, a) = T(5, b) = \{5\}$, and
- $T(6, a) = \{6\}$, $T(6, b) = \{3, 6\}$.

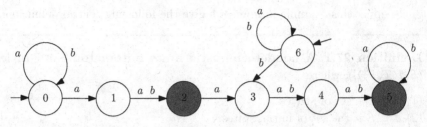

Fig. 27.2: Another example of an NFA

27.2   Runs of nondeterministic finite automata

Let $\mathcal{A} = (S, I, T, F, \Sigma)$ be an NFA and w be a string $\sigma_1\sigma_2\ldots\sigma_n$. How does the automaton $\mathcal{A}$ run on the input string w? We define this below and then explain the definition in more detail.

Definition 27.2. A **run** of the automaton $\mathcal{A}$ on the string $\sigma_1\sigma_2\ldots\sigma_n$ is a sequence of states

$$s_1, s_2, \ldots, s_n, s_{n+1}$$

such that $s_1 \in I$ and $s_{i+1} \in T(s_i, \sigma_i)$ for all $i = 1, \ldots, n$.

A run of $\mathcal{A}$ on $w = \sigma_1\ldots\sigma_n$ can be viewed as an execution of the algorithm $Run(\mathcal{A}, w)$, shown in Algorithm 27.1.

Algorithm 27.1 $Run(\mathcal{A}, w)$ algorithm

(1) Initialize $i = 1$ and randomly select an $s \in I$.
(2) Print s.
(3) *While $i \leq n$ do*
    (a) Set $\sigma = \sigma_i$.
    (b) Randomly select a state q from $T(s, \sigma)$.
    (c) Set $s = q$ and print s.
    (d) Increment i.

This algorithm outputs the states of the DFA $\mathcal{M}$ as it reads through the string w. First, the algorithm prints one of the initial states. It then reads the first input symbol σ_1, makes a random σ_1-transition from the initial state to one of the states q in $T(q_0, \sigma_1)$, prints q, reads the next symbol σ_2, makes a random σ_2-transition from q, and so on. An important thing to note is that there could be *several runs* of the automaton on one input w.

A second way of looking at how $\mathcal{A}$ processes an input string $w = \sigma_1 \sigma_2 \ldots \sigma_n$ is this. Suppose $\mathcal{A}$ has k initial states q_1, ..., q_k. Before the automaton starts reading w, the automaton splits itself into k copies where the ith copy is at state q_i, $i = 1, \ldots, k$. Each copy of $\mathcal{A}$ now reads the first symbol σ_1 of w. Assume that one of these copies is at state q, and there are t transitions from q labeled with σ_1. This copy now splits itself into t new copies where each moves along one of the σ_1-transitions. This continues as before. Thus, at any given stage of the computation there are several copies of the automaton reading the same symbol in *parallel*. These copies all make independent transitions by splitting themselves and thus making more copies if that is dictated by the transition function. If a copy of the automaton is in a state where there are *no* transitions, then the copy dies.

Yet another helpful and more formal way to think about how $\mathcal{A}$ processes the input string $w = \sigma_1 \ldots \sigma_n$ is to represent all the runs as a tree. The root of the tree is the start of the computation. Every branching node of the tree represents the fact that $\mathcal{A}$ splits itself at that node by reading the next symbol from w. The height of the tree is at most $n + 1$. The paths of length $n + 1$ from the root to the leaves correspond to the runs of the automaton on w.

For example, consider the automaton $\mathcal{A}$ represented in Figure 28.1. For this automaton, let the input string w be *ababaab*. Figure 27.3 represents all the runs of $\mathcal{A}$ on w as a tree. In this figure, we can see that the two rightmost paths die out because at state 3 the automaton can not make any transitions, and therefore along these two paths the automaton could not process the rest of the input string w. The other paths produce runs of the automaton on the input. One of these runs is such that the last state of the run is the accepting state. Here is a formal definition of acceptance:

Definition 27.3. We say that a NFA $\mathcal{A}$ **accepts** a string $w = \sigma_1 \sigma_2 \ldots \sigma_n$ if the automaton has at least one run $s_1, s_2, \ldots, s_{n+1}$ on w such that the state s_{n+1} is an accepting state. In this case we say that the run $s_1, s_2, \ldots, s_{n+1}$ is an **accepting run**. Thus, the automaton $\mathcal{A}$ **does not accept** w if *no* run of $\mathcal{A}$ on w is accepting.

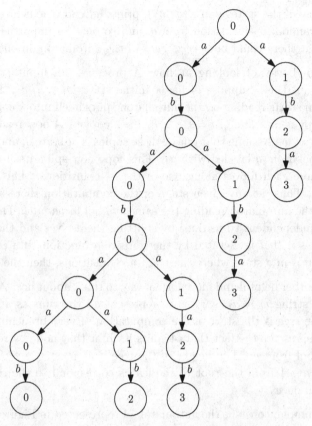

Fig. 27.3: A tree representation of a computation

Informally, $\mathcal{A}$ accepts w if in the transition diagram of $\mathcal{A}$ there exists a path from an initial state to an accepting state such that the path is labeled with the string w. Thus, the acceptance of w by $\mathcal{A}$ is an *existential condition* because the condition refers to the existence of an accepting run of $\mathcal{A}$ on w.

Definition 27.4. Let $\mathcal{M} = (S, I, T, F, \Sigma)$ be an NFA. The **language accepted by** $\mathcal{M}$, denoted by $L(\mathcal{M})$, is the following language:

$$\{w \mid \text{the automaton } \mathcal{M} \text{ accepts } w\}.$$

A language $L \subseteq \Sigma^\star$ is **NFA recognizable** if there exists an NFA $\mathcal{M}$ such that $L = L(\mathcal{M})$.

Every DFA is an NFA. Therefore every DFA recognizable language is NFA recognizable. Now we give several simple examples of NFA recognizable languages.

Example 27.1. Consider the NFA with exactly one state with no transitions. If the state is accepting then the automaton accepts the language $\{\lambda\}$. Otherwise, the automaton accepts the empty language $\emptyset$.

Example 27.2. The automaton presented in Figure 28.1 recognizes the following language

$$L = \{u \in \{a, b\}^* \mid u \text{ has the symbol } a \text{ appearing in the third position from the right of } u\}.$$

Example 27.3. Consider the following automaton $\mathcal{M}$:

(1) The states are 0, 1, ..., 7.
(2) The initial state is 0, and the accepting states are 3 and 8.
(3) The transition T is defined as follows. $T(0, a) = \{0, 1\}$, $T(1, a) = T(1, b) = \{2\}$, $T(2, a) = T(2, b) = \{3\}$, $T(0, b) = \{0, 4\}$, $T(4, a) = T(4, b) = \{5\}$, $T(5, a) = T(5, b) = \{6\}$, and $T(6, a) = T(6, b) = \{7\}$.

This automaton accepts all the strings w of the alphabet $\{a, b\}$ such that w has either an a at the third position from the right or has a b at the fourth position from the right of w. The reader can easily draw the transition diagram of this NFA. endenumerate

27.3 Exercises

(1) For each language described below, draw the transition diagram of the NFA which recognizes it. The NFA must have exactly the specified number of states. The alphabet is $\{a, b\}$.

(a) $\{w \mid w \text{ ends with } aaa\}$. (4 states)
(b) $\{w \mid w \text{ contains a substring } aub \text{ such that } u \text{ has length 6}\}$. (9 states)
(c) $\{w \mid w \text{ contains substring } abab\}$. (5 states)
(d) $\{w \mid w \text{ ends with an } a \text{ and no } a \text{ occurs between any occurrences of } b\}$ (3 states)

(2) We say that an NFA $\mathcal{A}$ is equivalent to another NFA $\mathcal{B}$ if $L(\mathcal{A}) = L(\mathcal{B})$. Prove that every NFA $\mathcal{A}$ is equivalent to an NFA with exactly one initial state.

(3) Is it true that every NFA $\mathcal{A}$ is equivalent to an NFA with exactly one initial state and exactly one final state? Justify your answer.

(4) Prove that if L_1 is recognized by an NFA with n states and L_2 is recognized by an NFA with m states, then $L_1 \cup L_2$ can be recognized by an NFA with $n + m$ states.

(5) Let W be a language. The language $Prefix(W)$ is the set of all strings that are prefixes of strings in W. Formally, $Prefix(W) = \{u \mid u$ is a prefix of a string from $W\}$. Prove that if W is FA recognizable, then so is $Prefix(W)$.

(6) For a language L, set $L^{-1} = \{w \mid w^{-1} \in L\}$, where w^{-1} is the *inverse* of w. For instance, for $w = aabbbab$ we have $w^{-1} = babbbaa$. Prove that L is NFA recognizable if and only if L^{-1} is NFA recognizable.

(7) For each natural number $n \geq 1$, consider the language

$$L_n = \{uav \mid u, v \in \{a, b\}^* \text{ and the length of } v \text{ is } n\}.$$

Prove that no DFA with less than 2^n states recognizes L_n.

Programming exercises

(1) Design an algorithm that given an NFA $\mathcal{A}$ and a string w, checks if $\mathcal{A}$ accepts w. Write a program that implements the algorithm.

(2) Design a program that given an NFA $\mathcal{A}$ does the following:

   (a) determines if $\mathcal{A}$ is a deterministic finite automaton,
   (b) outputs all the states in $\mathcal{A}$ that are not deterministic states, and
   (c) builds an equivalent NFA $\mathcal{B}$ with exactly one initial state and one accepting state.

Lecture 28

The subset construction

In the middle of difficulty lies opportunity.
Albert Einstein.

28.1 Converting NFA to DFA

Since every DFA recognizable language is NFA recognizable, a natural question to ask is whether every NFA recognizable language is DFA recognizable. In this lecture, we see that this is indeed the case. We design an algorithm that, given a nondeterministic finite automaton $\mathcal{A}$, produces a deterministic finite automaton $\mathcal{B}$ such that $L(\mathcal{A}) = L(\mathcal{B})$. So, let $w = \sigma_1 \sigma_2 \dots \sigma_n$ be an input for the NFA $\mathcal{A}$. We explain a process that will give us an idea of how to construct the desired DFA $\mathcal{B}$.

To start, we mark *every* initial state of $\mathcal{A}$. We will keep track of all the states that are currently marked. We read the first symbol σ_1 of w. For *every marked* state s we proceed as follows.

(1) Remove the mark on state s.
(2) Mark all states q if there is a transition from s to q labeled with σ_1.

It may be the case that s is marked again because there could be a transition from s or some other previously marked s' to s labeled with σ_1. The above process, after reading σ_1, marks several states of the automaton. The collection of states that are marked might now be different from the marked states we started with. We maintain the set of all states that are currently marked. We repeat the same process by reading the next symbol σ_2, then σ_3, and so on. The intention is that marked states are the states that the NFA might be at the given stage of the run. Thus, all we need to remember in order to process a given input signal σ is the set of states that are marked at any given time. In other words, the subsets of S are now playing the role of states of the new automaton that we want to build.

The above reasoning can be made more formal as follows. Let $w = \sigma_1 \ldots \sigma_n$ be a string and $\mathcal{A} = (S, I, T, F)$ be an NFA. Define the sequence $Q_1, \ldots, Q_{n+1}$ of subsets of S inductively as follows:

$$Q_1 = I, \ Q_2 = \bigcup_{p \in Q_1} T(p, \sigma_1), \ldots, \ Q_{n+1} = \bigcup_{p \in Q_n} T(p, \sigma_n).$$

Each Q_i is a subset of S. Now we reason as follows.

Assume that w is accepted by $\mathcal{A}$. Then, there exists an accepting run

$$q_1, \ldots, q_{n+1}$$

of $\mathcal{A}$ on w. Note that each q_i belongs to Q_i. Therefore, Q_{n+1} contains the accepting state q_{n+1}.

Consider the sequence

$$Q_1, \ Q_2, \ldots, \ Q_n, \ Q_{n+1}.$$

Note that for *every* state $q \in Q_{i+1}$, there exists a state $p \in Q_i$ such that $q \in T(p, \sigma_i)$, for all $i = 1, \ldots, n$. Therefore, for every state $q_{n+1} \in Q_n$ we can produce a sequence

$$q_1, \ q_2, \ \ldots, \ q_{n+1}$$

such that $q_{i+1} \in T(q_i, \sigma_i)$, for all $i = 1, \ldots, n$. This sequence is clearly a run of the automaton on the input w. Hence, if Q_{n+1} contains an accepting state then $\mathcal{A}$ must have an accepting run on w.

These observations guide us toward the following construction of a DFA $\mathcal{B} = (S', q_0', T', F')$:

(1) The set of states S' of $\mathcal{B}$ is $P(S)$.
(2) The initial state q_0' of $\mathcal{B}$ is I.
(3) The transition function T' is defined as follows. For any state $Q \in S'$ (that is a subset of S) and any symbol $\sigma \in \Sigma$,

$$T'(Q, \sigma) = \bigcup_{p \in Q} T(p, \sigma).$$

(4) The set F' contains all states $Q \in S'$ such that Q' has an accepting state of $\mathcal{A}$.

The automaton $\mathcal{B}$ is clearly a deterministic finite automaton. We have already shown that $L(\mathcal{A}) = L(\mathcal{B})$. Therefore, we have proved the following theorem. The theorem is an important result in theoretical computer science and computability.

Theorem 28.1 (The determinization theorem). *For any nondeterministic finite automaton $\mathcal{A}$, there exists a deterministic finite automaton $\mathcal{B}$ such that $\mathcal{A}$ and $\mathcal{B}$ are equivalent.*

The construction of the DFA from a given NFA $\mathcal{A}$ provided above is known as **subset construction**. Note that if the NFA $\mathcal{A}$ has n states then the produced DFA has 2^n states. Thus, there is an exponential blow up in the number of states of the input automaton $\mathcal{A}$. This blow up is inevitable by Exercise 7 in the previous lecture.

As an application of the determinization theorem, consider the language L discussed at the start of Lecture 27:

$$L = \{uav \mid u, v \in \{a, b\}^* \text{ and the length of } v \text{ is } n - 1\}.$$

There exists an NFA with n states that recognizes L. By the theorem above, L is also DFA recognizable.

Since NFA and DFA recognizability are equivalent notions (as we have proved), we often use the term *finite automata (FA) recognizable* to mean either and both of these notions.

28.2   NFA with silent moves

An NFA with silent moves is an NFA equipped with λ-transitions (recall that λ represents the empty string). The meaning of λ-transitions is the following. Assume that the machine is in state s and decides to move along a λ-transition from state s to a state q. The machine, instead of reading the next symbol σ, moves to q and thus changes its state from s to q. This move of the machine is sometimes called a *silent move* or a *silent transition* since the next symbol of the input string has not been processed. In state q, the machine can either make a σ-transition or again repeat its silent move (if there is a silent transition from q). In the first case, the machine acts as an NFA, that is it changes its state and reads the symbol and moves to the next symbol of the input string. In the second case, the machine makes a silent move by changing its state only. Informally, silent moves are internal transitions of the machine. With these transitions, the machine does not process the input signal but rather prepares itself to process the signal. Formally, an NFA with silent moves is defined as follows:

Definition 28.1. Let Σ be an alphabet. Denote by $\Sigma_\lambda = \Sigma \cup \{\lambda\}$. An **NFA with silent moves** is a tuple $(S, I, T, F, \Sigma_\lambda)$ such that:

(1) S is the finite set of states of $\mathcal{A}$,
(2) $I \subseteq S$ is the set of initial states,
(3) T is a transition function $T : S \times \Sigma_\lambda \to P(S)$, and
(4) $F \subseteq S$ is the set of accepting states.

On input $w = \sigma_1 \ldots \sigma_n \in \Sigma^*$, the automaton $\mathcal{A}$ *runs* as follows. The automaton $\mathcal{A}$ nondeterministically selects an initial state, call it q_1. At state q_1, the automaton has the following two options:

(1) Select nondeterministically a transition from q_1 labeled with σ_1.
(2) Select nondeterministically a silent transition without reading the symbol σ_1.

In the first case, the automaton is ready to start processing the next input symbol σ_2. In the second case, the automaton still needs to process the first symbol σ_1. This continues on. Assume that the automaton has arrived to state s and has processed the $\sigma_1 \ldots \sigma_{i-1}$ part of the input string. At state s, the automaton has two possibilities as above:

(1) Select nondeterministically a transition from s labeled with σ_i.
(2) Select nondeterministically a silent transition without reading the symbol σ_i.

Again, in the first case, $\mathcal{A}$ is now ready for processing the next input symbol σ_{i+1}. In the second case, the automaton still needs to process the first symbol σ_i. Once the entire string w is processed, the automaton can still continue on running using its silent transitions. We say that the run is accepting if the automaton can enter an accepting state after processing the entire string w. In this case, we say that $\mathcal{A}$ **accepts** the string w.

From the definition, it seems that NFA with silent moves are a lot more general than NFA without them. Our goal is to show that every nondeterministic automaton $\mathcal{A} = (S, I, T, F)$ with silent moves can be simulated by an NFA without silent moves. In other words, we want to construct an NFA $\mathcal{B}$ such that $L(\mathcal{A}) = L(\mathcal{B})$. Let s be a state of $\mathcal{A}$. Define $SilentMoves(s)$ recursively as follows:

Base case. $SilentMoves_0(s) = \{s\}$.

Inductive Step $k + 1$. Assume we have defined $SilentMoves_k(s)$. Set

$$SilentMoves_{k+1}(s) = SilentMoves_k(s) \cup \{q \mid \text{there is a silent transition} \\ \text{from some } p \in SilentMoves_k(s) \text{ to } q\}.$$

Define

$$SilentMoves(s) = SilentMoves_0(s) \cup SilentMoves_1(s) \cup \ldots \cup SilentMoves_n(s),$$

where n is the number of states in the automaton $\mathcal{A}$. Thus, $SilentMoves(s)$ is the collection of all states that are reachable from the state s using the λ-transitions. If the automaton is in state s and the next symbol is σ, then the automaton, by moving along the λ-transitions, can select any of the states q in $SilentMove(s)$ and move along a transition labeled by σ from q. Therefore, we can modify the transition function T of $\mathcal{A}$ as follows:

$$T'(s, \sigma) = \bigcup_{q \in SilentMoves(s)} T(q, \sigma).$$

We now construct the following automaton $\mathcal{B} = (S', I', T', F')$:

(1) $S' = S$.
(2) $I' = I$.
(3) The transition function T' is defined above.
(4) $F' = \{q \mid SilentMoves(q) \cap F \neq \emptyset\}$.

Clearly $\mathcal{B}$ is an NFA that does not have λ-transitions. We want to show that the automaton $\mathcal{A}$ with silent moves and the automaton $\mathcal{B}$ without silent moves are equivalent.

Assume that $w = \sigma_1 \ldots \sigma_n$ is accepted by the NFA $\mathcal{B}$. We want to show that the automaton $\mathcal{A}$ accepts w. Take an accepting run

$$s_1, \ldots, s_{n+1}$$

of $\mathcal{B}$ on w. Then, by the definition of I', s_1 is an initial state of $\mathcal{A}$. By the definition of T', the automaton $\mathcal{A}$ makes zero or more silent transitions from state s_1 to a state s'_1 at which $\mathcal{A}$ selects a σ_1-transition to state s_2. Again, using the definition of T', from state s_2 the automaton $\mathcal{A}$ makes zero or more silent transitions to state s_2' at which $\mathcal{A}$ selects a σ_2-transition to state s_3. This continues on. Thus, we can build a run of the original automaton $\mathcal{A}$:

$$s_1, \ldots, s_1', s_2, \ldots, s_2', s_3, \ldots, s_3', s_4, \ldots s_4', \ldots, s_{n-1}', s_n, \ldots, s_n', s_{n+1}.$$

Since s_{n+1} is an accepting state of $\mathcal{B}$, this means $\mathcal{A}$ can make 0 or more silent transitions to reach the accepting state s_{n+1} in $\mathcal{A}$. This shows that $\mathcal{A}$ accepts the string w.

Assume now that $w = \sigma_1 \ldots \sigma_n$ is accepted by the NFA $\mathcal{A}$. Take an accepting run

$$(\star) \qquad s_1, \ldots, s_1', s_2, \ldots, s_2', \ldots, s_{n-1}, \ldots, s_{n-1}', s_n, \ldots, s_n', s_{n+1}, \ldots, s_{n+1}'.$$

of $\mathcal{A}$ on w. For this run we have the following properties:

(1) s_1 is an initial state of $\mathcal{A}$.
(2) For each $i = 1, \ldots, n$, $s_{i+1} \in T(s_i', \sigma_i)$.
(3) For each $i = 1, \ldots, n, n+1$, the part $s_i, \ldots, s_i'$ of the run in $(\star)$ is due to silent moves of the automaton $\mathcal{A}$.
(4) $s_{n+1}' \in F$.

From the properties above and the construction of $\mathcal{B}$ we have the following. The state s_1 is an initial state of $\mathcal{B}$. For each i, $s_{i+1} \in T'(s_i, \sigma_i)$. The state $s_{n+1} \in F'$. Thus, $\mathcal{B}$ has an accepting run on w. Therefore $L(\mathcal{A}) = L(\mathcal{B})$.

Thus, we have shown that NFA with silent moves are equivalent to NFA without silent moves. Gathering the results of this and the previous sections, we can now state the following theorem:

Theorem 28.2. *Let L be a language. The following statements are equivalent:*

(1) There exists a DFA recognizing L.
(2) There exists an NFA recognizing L.
(3) There exists an NFA with silent moves recognizing L. □

28.3   Exercises

(1) Consider the language
$$L = \{uav \mid u, v \in \{a, b\}^* \text{ and the length of } v \text{ is } 2\}.$$
An NFA $\mathcal{A}$ recognizing this language is shown in Figure 28.1. Apply the subset construction to $\mathcal{A}$ and draw a transition diagram of the resulting deterministic automaton.

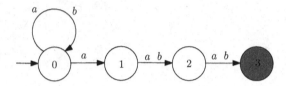

Fig. 28.1: An example of an NFA

(2) Let $\mathcal{A} = (S, I, T, F)$ be an NFA. Construct a new set S' by induction as follows.

- **Base case.** Set $S_0 = \{I\}$.
- **Inductive step** $n + 1$. Suppose that S_n has been built and is $\{X_1, \ldots, X_k\}$. Set
$$S_{n+1} = S_n \bigcup \{T'(X_i, \sigma) \mid \sigma \in \Sigma, \ 1 \le i \le k\},$$
where $T'(Q, \sigma) = \bigcup_{p \in Q} T(p, \sigma)$.

Set $S' = S_k$, where $S_k = S_{k+1}$. Prove the following two facts.

(a) There exists a k such that $S_{k+1} = S_k$.

(b) Set $F' = \{X \mid X \in S', X \cap F \ne \emptyset\}$. Define $\mathcal{A}' = (S', \{I\}, T', F')$. Show that $\mathcal{A}'$ is a deterministic automaton that is equivalent to $\mathcal{A}$.

(3) Prove that if W is FA recognizable, then so is W^*.

(4) Prove that if W and U are FA recognizable, then so is $W \cdot U$.

Programming exercises

(1) Write a program *Determinize($\mathcal{A}$)* that given an NFA $\mathcal{A}$, outputs a DFA $\mathcal{B}$ equivalent to $\mathcal{A}$.

(2) Design an algorithm that given an NFA $\mathcal{A}$ (which may be with silent moves) and a string w, checks if $\mathcal{A}$ accepts w. Write a program that implements the algorithm.

(3) Write a program $Transform(\mathcal{A})$ that given an NFA $\mathcal{A}$ with silent moves, outputs an NFA $\mathcal{B}$ (without silent moves) equivalent to $\mathcal{A}$.

Lecture 29

Regular expressions and finite automata

*The more I think about language, the more it amazes
me that people ever understand each other at all.*

Kurt Gödel.

29.1 Regular expressions

Regular expressions are used in many programming languages and tools.
They can be applied to find and extract patterns in texts and programs. For
example, using regular expressions, we can find lines in texts that contain
specific words, characters or letters. We can replace, modify, clean and
reformat sections in html and xml documents. Using regular expressions,
we can also specify and validate formats of data such as passwords, e-mail
addresses, user logins, and so on. The goal of this section is to introduce
regular expressions and study their relationship with finite automata. In
particular, we will describe methods to convert regular expressions into
finite automata, and finite automata into regular expressions. We start
with the definition of regular expression.

Definition 29.1. Let Σ be an alphabet. We define **regular expressions**
over the alphabet Σ by induction.
Base Case: The following are regular expressions: each letter σ of Σ, and
the symbols $\emptyset$ and λ.
Inductive step: Assume that r_1 and r_2 are regular expressions that have
already been defined. Then each of the following is a regular expression:
(1) $(r_1 + r_2)$, (2) $(r_1 \cdot r_2)$, and (3) (r_1^*).

Thus, every regular expression is a string that is built using the rules above. For example, for the alphabet $\{a, b\}$, the following are regular expressions:

$$a, \ b, \ ((a^\star) + (b^\star)), \ (((a^\star) + (b^\star)) \cdot (a \cdot b)), \ (((a^\star) + (b^\star)) + \lambda).$$

To represent regular expressions, we use symbols r, s, t, etc. Note that a regular expression is simply a string over the alphabet $\Sigma \cup \{ +, \cdot, \star,), (, \lambda, \emptyset \}$. Clearly, not all strings over this alphabet are regular expressions.

In the lecture on trees (Lecture 9), we defined arithmetic expressions. Just like regular expressions, arithmetic expressions are strings that are built using some simple rules. Every arithmetic expression without an integer variable can be *evaluated*. The values of such expressions are integers. For example, the value of the arithmetic expression $(((1+7)-(25-19)) \times 2)$ is 4. Just like evaluating arithmetic expressions, we can also evaluate regular expressions. The values of the regular expressions are languages.

Let r be a regular expression. The regular expression r can be evaluated to a language, denoted by $L(r)$. We call $L(r)$ **the language of** r. Since r is defined inductively, the language $L(r)$ is also defined inductively. The reader needs to recall the concatenation and the star operations on languages introduced in Lecture 25.

Definition 29.2. For the regular expression r, the **language of** r, denoted by $L(r)$, is defined as follows:

Base Case:

(1) If $r = a$, where $a \in \Sigma$, then $L(r) = \{a\}$.
(2) If $r = \emptyset$ then $L(r) = \emptyset$.
(3) If $r = \lambda$ then $L(r) = \{\lambda\}$.

Inductive step: Assume that the languages $L(r_1)$ and $L(r_2)$, for regular expressions r_1 and r_2, have been defined. Then:

(1) $L(r_1 + r_2) = L(r_1) \cup L(r_2)$.
(2) $L(r_1 \cdot r_2) = L(r_1) \cdot L(r_2)$.
(3) $L(r_1^\star) = L(r_1)^\star$.

Note that in the inductive step, we use the symbols $\cdot$ and $\star$ in two different ways. On the left side of the equalities, the symbols $\cdot$ and $\star$ are used syntactically (as constructors for regular expressions). On the right side of the equalities, the symbols are interpreted as the concatenation and star operations on the languages. The symbol $+$ is interpreted as the union operation. Similarly, in the base case, the symbol $\emptyset$ is used syntactically, and on the right side it is used to represent the empty set.

We now present several examples of languages $L(r)$ defined by regular expressions. In all these examples, the alphabet is $\{a, b\}$.

(1) $L(a \cdot b + b) = \{ab, b\}$.
(2) $L(b^\star) = \{\lambda, b, bb, bbb, bbbb, \ldots\}$.
(3) $L(a^\star \cdot b^\star) = \{a^n b^m \mid n \geq 0 \text{ and } m \geq 0\}$.
(4) $L((a + b)^\star \cdot (a \cdot b \cdot a) \cdot (a + b)^\star) = \{w \mid w \text{ contains the substring } aba\}$.
(5) $L(\emptyset^\star) = \{\lambda\}$.
(6) $L(a^\star \cdot a \cdot (bbbaaa) \cdot b \cdot b^\star) = \{w \mid \text{there exist nonempty strings } u \text{ and } v$ such that $w = uabbbaaabv$, where u consists of a's only, and v consists of b's only$\}$.

Strictly speaking, in our examples above we did not follow the syntax of writing the regular expressions. For example, by $a \cdot b + b$ we meant the regular expression $((a \cdot b) + b)$. Thus, we sometimes omit parentheses and the concatenation symbol $\cdot$ to prevent our notation from becoming too cumbersome. The general rules say that $\cdot$ is of higher precedence than $+$, and $\star$ is higher precedence than $\cdot$. Thus, for example $a \cdot b + a \cdot b^\star$ represents $((a \cdot b) + (a \cdot (b^\star)))$.

Definition 29.3. We say that a language $L \subseteq \Sigma^\star$ is **regular** if there exists a regular expression r such that $L = L(r)$. In this case, we also say that r **represents** the language L.

Note that if r represents L then $r + \emptyset$, $r + \emptyset + \emptyset$, ... all represent L. Therefore, every regular language has infinitely many regular expressions representing it. Now observe that we have seen two quite different ways of describing languages. The first uses finite automata (as in the previous lectures), and the second uses regular expressions. In the next two sections, we show that these two methods of describing languages are actually equivalent.

29.2 From regular expressions to finite automata

The goal of this section is to provide a method to convert a regular expression r into an automaton recognizing $L(r)$. Here is the main result of this section:

Theorem 29.1. *Every regular language is FA recognizable.*

Proof.

Let r be a regular expression. We want to prove that $L(r)$ is FA recognizable. In our proof, we follow the inductive definition for the language $L(r)$, given in the previous section. Our proof uses NFA with silent moves.

Base Case: Assume that r is defined in the base case. Then, $L(r)$ is either $\emptyset$, $\{\lambda\}$, or $\{a\}$ for some $a \in \Sigma$. Clearly, in each case, $L(r)$ is FA recognizable.

Inductive step: In this step there are three cases to consider.

Case 1. Assume that r is of the form $(r_1 + r_2)$. We now use the inductive hypothesis applied to r_1 and r_2. By the hypothesis, the languages $L(r_1)$ and $L(r_2)$ are FA recognizable. The language $L(r_1 + r_2)$ is by definition $L(r_1) \cup L(r_2)$. We know that the union of FA recognizable languages is again FA recognizable. Hence, $L(r)$ is FA recognizable.

Case 2. Assume that r is of the form $(r_1 \cdot r_2)$. By the inductive hypothesis, the languages $L(r_1)$ and $L(r_2)$ are FA recognizable. The language $L(r_1 \cdot r_2)$ is by definition $L(r_1) \cdot L(r_2)$. We want to show that $L(r_1) \cdot L(r_2)$ is FA recognizable.

Let $\mathcal{A}_1 = (S_1, I_1, T_1, F_1)$ and $\mathcal{A}_2 = (S_2, I_2, T_2, F_2)$ be finite automata recognizing $L(r_1)$ and $L(r_2)$, respectively. Both automata are NFA that may have silent moves. We may assume that the state sets S_1 and S_2 have no states in common. Using these two automata, we want to build a finite automaton recognizing $L(r_1) \cdot L(r_2)$.

A machine that accepts strings in $L(r_1) \cdot L(r_2)$ can be described as follows. Suppose w is an input string. We simulate the machine $\mathcal{A}_1$ and run $\mathcal{A}_1$ on w. If we never reach an accepting state of $\mathcal{A}_1$, then we reject w. Assume, that we have reached an accepting state after processing some prefix v of w. In this case, we nondeterministically choose one of the following actions: (1) continue running $\mathcal{A}_1$ on w, or (2) make a silent transition to the initial state of $\mathcal{A}_2$ and start simulating $\mathcal{A}_2$.

This description suggests that we can put the automata $\mathcal{A}_1$ and $\mathcal{A}_2$ together as follows. We keep the states and transition tables of both machines. We declare the initial states of $\mathcal{A}_1$ to be the initial states, and the accepting states of $\mathcal{A}_2$ to be the accepting states of our machine. Finally, we add the λ-transitions from the accepting states of $\mathcal{A}_1$ to the initial states of $\mathcal{A}_2$. Based on this description, we formally define the automaton $\mathcal{A} = (S, I, T, F)$ recognizing $L(r_1) \cdot L(r_2)$ as follows:

(1) $S = S_1 \cup S_2$.
(2) $I = I_1$ and $F = F_2$.
(3) For state $s \in S$ and input symbol $\sigma \in \Sigma \cup \{\lambda\}$, the transition function T is defined according to the following rules.

    (a) If $s \in S_1$ and s is not an accepting state then $T(s, \sigma) = T_1(s, \sigma)$.
    (b) If $s \in S_1$, s is an accepting state and $\sigma \neq \lambda$, then $T(s, \sigma) = T_1(s, \sigma)$.
    (c) If $s \in S_1$, s is an accepting state and $\sigma = \lambda$, then $T(s, \sigma) = T_1(s, \sigma) \cup I_2$.
    (d) If $s \in S_2$ then $T(s, \sigma) = T_2(s, \sigma)$.

It is not hard to check that $\mathcal{A}$ recognizes $L(r_1) \cdot L(r_2)$.

Case 3. Assume that r is of the form $(r_1)^*$. By the inductive hypothesis, the language $L(r_1)$ is FA recognizable. The language $L(r_1^*)$ is by definition $L(r_1)^*$. We want to show that $L(r_1)^*$ is FA recognizable.

Let $\mathcal{A}_1 = (S_1, q_1, T_1, F_1)$ be a finite automaton recognizing $L(r_1)$. We want to construct a finite automaton that recognizes the language $L(r_1)^*$. For an input string w, we process w as follows. If w is the empty string, then we accept it because the star of every language contains λ. Otherwise, we simulate $\mathcal{A}_1$ and read w. Every time we reach an accepting state of $\mathcal{A}_1$, we have to make a nondeterministic choice. We either continue on running $\mathcal{A}_1$ or make a silent move to one of the initial states of $\mathcal{A}_1$. Thus, we construct our automaton $\mathcal{A}$ recognizing $L(r_1)^*$ as follows. We keep all the states and transitions of $\mathcal{A}_1$. We add λ-transitions from each accepting state of $\mathcal{A}_1$ to every initial state of $\mathcal{A}_1$. Finally, we add one new initial state q_{new} with no outgoing or ingoing transitions and declare it to be an accepting state. This is needed to accept the empty string. A more formal definition of the automaton $\mathcal{A} = (S, I, T, F)$ recognizing $L(r_1)^*$ is this:

(1) $S = S_1 \cup \{q_{new}\}$, where q_{new} is a state not in S_1.
(2) $I = I_1 \cup \{q_{new}\}$ and $F = F_1 \cup \{q_{new}\}$.

(3) For state $s \in S$ and input symbol $\sigma \in \Sigma \cup \{\lambda\}$, the transition function T is defined as follows:

   (a) If $s \in S_1$ and s is not an accepting state, then $T(s, \sigma) = T_1(s, \sigma)$.
   (b) If $s \in S_1$, s is an accepting state, and $\sigma \neq \lambda$, then $T(s, \sigma) = T_1(s, \sigma)$.
   (c) If $s \in S_1$, s is an accepting state, and $\sigma = \lambda$, then $T(s, \sigma) = T_1(s, \sigma) \cup I_1$.
   (d) If $s = q_{new}$, then $T(s, \sigma) = \emptyset$.

The automaton $\mathcal{A}$ recognizes $L(r_1)^{\star}$. We have proved the theorem. $\square$

29.3   Generalized finite automata

The goal of this section is to show how to convert a finite automaton $\mathcal{A}$ into a regular expression r such that $L(\mathcal{A}) = L(r)$. There are several methods to do this. Here, we present one method based on an elegant generalization of finite automata.

One can generalize finite automata in many ways. One generalization can be obtained by assuming that a machine processes an input not just once, but may, in fact, move its head back and forth as many times as needed. The machine can also write on tape. This produces the Turing machine model of computations. Turing machine model subsumes the finite automata model of computation. Namely, all NFA recognizable languages are recognized by some Turing machines, and there are languages recognized by Turing machines but not by finite automata. Another generalization of finite automata is obtained by providing an extra memory location for a machine. When the machine reads an input of length n, the memory location can store at most n amount of information. The machine is not allowed to re-read any symbol it has read so far. Such an automaton is called a pushdown automaton. The class of pushdown automata also subsumes finite automata model of computation but it is properly contained in the class of Turing machines. These both, Turing machines and pushdown automata, are not in the scope of this textbook. Instead, in this section, we explain machines called generalized finite automata that process input strings block by block, instead of symbol by symbol.

Informally, a generalized finite automaton is the same as a finite automaton except the labels of the transitions are now regular expressions. Thus, a generalized finite automaton has a finite set of states, some states are initial states, and some are accepting. In addition, there are transitions between the states labeled with regular expressions. Clearly, every NFA (with silent moves) is a generalized finite automaton because the labels of all its transitions are symbols from the alphabet or the empty string λ, and thus, are regular expressions. Here is a formal definition.

Definition 29.4. A **generalized nondeterministic finite automaton**, written GNFA for short, is a 5-tuple $\mathcal{A} = (S, I, T, F, \Sigma)$, where

(1) S is a finite, nonempty set called the **set of states**,
(2) I is a nonempty subset of S called the **set of initial states**,
(3) T is a function that labels transitions with regular expressions from the alphabet Σ,
(4) F is a subset of S called the **set of accepting states**, and
(5) Σ is a finite alphabet.

An example of a GNFA is in Figure 29.1.

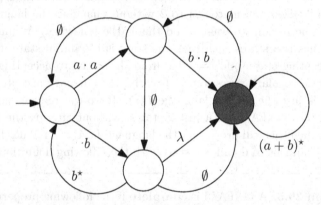

Fig. 29.1: A generalized nondeterministic finite automaton

A generalized finite automaton works as follows. The automaton reads a block of symbols from the input and not necessarily just one input symbol, then makes a transition along an edge. The block of symbols read must belong to the language described by the regular expression that labels the edge. Of course, the automaton is nondeterministic so it has several ways to process the block. The sequence of blocks is accepted by the automaton if the automaton can reach an accepting state by processing the sequence.

We say that a GNFA $\mathcal{A}$ **accepts** a string $w \in \Sigma^\star$ if for some $w_1, \ldots, w_n \in \Sigma^\star$ we have the following properties:

(1) $w = w_1 \ldots w_n$.
(2) There exists a sequence $s_1, s_2, \ldots s_{n+1}$ of states of $\mathcal{A}$ such that:
   (a) s_1 is an initial state,
   (b) s_{n+1} is an accepting state, and
   (c) For every i, $1 \leq i \leq n$, the word w_i belongs to the language $L(r)$, where the transition from s_i to s_{i+1} is labeled by r.

For example, the following strings are accepted by the GNFA given in Figure 29.1. *aabbbaba*, *bbbbbbbbaabb*, *b*, *aabb*, *baaaaaababababa*. The language recognized by a given GNFA $\mathcal{A}$ is the set of all strings accepted by $\mathcal{A}$. We use the same notation as for finite automata:

$$L(\mathcal{A}) = \{w \in \Sigma^\star \mid \mathcal{A} \text{ accepts } w\}.$$

For example, the language recognized by the GNFA in Figure 29.1 is this:

$$L = \{aabbv \mid v \in \{a, b\}^\star\} \cup \{bv \mid v \in \{a, b\}^\star\}.$$

Our goal is to show that it is possible to reduce the number of states of the generalized automaton $\mathcal{A}$ to 2 without changing the language recognized. The reason we want to do this is the following. Assume that a GNFA $\mathcal{B}$ has two states such that one state is the initial state, the other is the accepting state, and the edge from the initial to accepting state is labeled by a regular expression r. Clearly, the language accepted by the GNFA $\mathcal{B}$ is just the regular language $L(r)$. If we can reduce the original GNFA $\mathcal{A}$ to a GNFA $\mathcal{B}$ with just 2 states, without altering the language accepted, then this will prove that the language $L(\mathcal{A})$ is a regular language.

To explain this in detail, we start with the following definition:

Definition 29.5. A GNFA $\mathcal{A}$ is **complete** if the following properties hold:

(1) $\mathcal{A}$ has exactly one start state q_0.
(2) There exists exactly one accepting state f, and it is not equal to the start state.
(3) There is no transition from any state s to q_0.
(4) There is no transition from f to any state s.
(5) For each pair s, s' of states, where $s, s' \notin \{q_0, f\}$, there exists a transition from s to s'.

The lemma below shows that we can restrict ourselves to complete GNFA.

Lemma 29.1. *For any GNFA $\mathcal{A}$, there exists a complete GNFA $\mathcal{B}$ such that $\mathcal{A}$ and $\mathcal{B}$ accept the same language.*

Proof. We change the automaton $\mathcal{A}$ step-by-step as follows:

(1) We add a new initial state q_0 with a λ-transition to each of the original initial states, and declare the original initial states to no longer be initial.
(2) We add a new accepting state f with a λ-transition from each of the original accepting states to f, and declare original accepting states to no longer be accepting.
(3) For all $s, s' \notin \{s_0, f\}$, if there is no transition from s to s', then we add a transition from s to s' and label it with $\emptyset$.

Steps (1) and (2) guarantee the first four conditions needed for a GNFA to be complete. Step (3) guarantees condition 5) of the complete GNFA definition. None of the above three steps alter the language recognized. The resulting GNFA is equivalent to the original one. $\square$

We are now ready to prove the main theorem:

Theorem 29.2. *For every GNFA $\mathcal{A} = (S, I, T, F)$ we can construct a complete GNFA $\mathcal{B}$ with exactly two states such that $L(\mathcal{A}) = L(\mathcal{B})$.*

Proof.
We can assume that the automaton $\mathcal{A}$ is complete by the lemma above. The idea of our proof is this. We select a state s in $\mathcal{A}$, which is neither initial nor accepting. We remove the state and then repair the "damage" done, by relabeling and adding some transitions to the original automaton $\mathcal{A}$. These new transitions recover all the runs lost when the state s is removed. The new automaton has fewer states than the original automaton and accepts the same language. We continue this procedure until two states, the initial state and the accepting state, remain.

Now we present a formal construction, called $Reduce(\mathcal{A})$, of the desired automaton. Assume that $\mathcal{A}$ has exactly n states. Now proceed as follows:

(1) If $n = 2$, then $\mathcal{A}$ is the desired GNFA.

(2) Suppose that $n > 2$. Select a state s distinct from both the initial state and the accepting state. Construct the following GNFA $\mathcal{A}_1 = (S_1, I_1, T_1, F_1)$ where:

    (a) $S_1 = S \setminus \{s\}$.

    (b) Do not change the initial and accepting states.

    (c) For all $s_1, s_2 \in S$ that are distinct from s such that $T(s_1, s) = r_1$, $T(s, s) = r_2$, $T(s, s_2) = r_3$, $T(s_1, s_2) = r_4$ label the edge from s_1 to s_2 with the regular expression $r_1 \cdot r_2^* \cdot r_3 + r_4$. Thus, $T_1(s_1, s_2) = r_1 \cdot r_2^* \cdot r_3 + r_4$.

Figure 29.2 depicts how the "damage" is repaired when we remove state s from a part of the transition diagram.

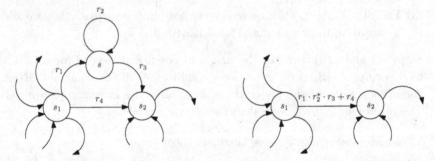

The "damage" will be done by removing the state s.   The "damage" is repaired.

Fig. 29.2: Repairing the "damage"

The constructed automaton $\mathcal{A}_1$ has the following properties. The number of states of $\mathcal{A}_1$ is $n-1$ and the automaton is still complete. Furthermore, $\mathcal{A}_1$ recognizes exactly the same language as $\mathcal{A}$. Indeed, if $\mathcal{A}$ makes a transition without going through state s, then the transition is unaffected, and thus is also a transition of $\mathcal{A}_1$. If $\mathcal{A}$ makes a transition from state s_1 to s, then from s to s_2, where s_1 and s_2 are distinct from s, then these two transitions of $\mathcal{A}$ can be replaced by one transition from s_1 to s_2 in $\mathcal{A}_1$, due to the definition of T_1. Similarly, any transition in $\mathcal{A}_1$ can be simulated by transitions in $\mathcal{A}$.

In order to complete the construction of the desired automaton, we continue the procedure above until two states, the initial and accepting states, are left.   □

Corollary 29.1. *Every FA recognizable language L is regular.*

Proof. Let $\mathcal{A}$ be an NFA that recognizes L. This automaton is also a GNFA. By the theorem above, there exists a complete two-state GNFA $\mathcal{B} = (S, I, T, F)$ such that $L(\mathcal{B}) = L$. Let s_0 be the initial state and f be the accepting state of $\mathcal{B}$. Let $r = T(s_0, f)$. Then $L = L(r)$. □

Thus, we have proved the following theorem known as Kleene's theorem.

Theorem 29.3 (Kleene's theorem). *A language L is FA recognizable if and only if L is regular.*

29.4 Exercises

(1) Write down regular expressions representing the following languages over the alphabet $\{a, b\}$:

    (a) $\{wa \mid w \in \Sigma^*\}$.
    (b) $\{\lambda\}$.
    (c) $\{w \mid w \in \Sigma^* \text{ and } w \neq \lambda\}$.
    (d) $\{uaabv \mid u, v \in \Sigma^*\}$.
    (e) $\{w \mid w \text{ ends with } aaa\}$.
    (f) $\{w \mid w \text{ contains a substring } aub \text{ such that } u \text{ has length 6}\}$.
    (g) $\{w \mid w \text{ contains substring } abab\}$.
    (h) $\{w \mid w \text{ ends with } a \text{ and no } a \text{ occurs between any occurrences of } b\}$.
    (i) $\{w \mid w \text{ has } a \text{ at the 9th position from the right of } w\}$.

(2) Prove that every finite language is regular.

(3) Design an algorithm that checks if a string r is a regular expression of the alphabet $\{a, b\}$.

(4) Prove that the automaton constructed in Case 2 of Theorem 29.1 recognizes the language $L(r_1) \cdot L(r_2)$.

(5) Prove that the automaton constructed in Case 3 of Theorem 29.1 recognizes the language $L(r_1)^*$.

(6) For each of the following regular expressions r, construct finite automata recognizing the language $L(r)$:

    (a) a.
    (b) a^*.
    (c) $(a^* + b^*)$.
    (d) $(a \cdot b)$.
    (e) $(a \cdot b)^*$.

(f) $(a^\star + b^\star) \cdot (a \cdot b)^\star$.

Your construction should follow the proof of Theorem 29.1.

(7) Consider the automaton $\mathcal{A}$ in Figure 29.3 below.

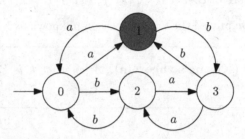

Fig. 29.3: An automaton for reduction

Do the following:

(a) Transform the automaton to a complete automaton $\mathcal{B}$.
(b) Apply the $Reduce(\mathcal{A})$ algorithm to the automaton, step-by-step, to get a regular expression for $L(\mathcal{A})$. Draw transition diagrams of the resulting generalized nondeterministic finite automata at each stage.

Programming exercises

(1) Write a program that checks if a string r is a regular expression of the alphabet $\{a, b\}$.
(2) Implement the $Reduce(\mathcal{A})$ algorithm.
(3) Write a method that given a regular expression r, constructs an automaton A that recognizes $L(r)$.

Algorithms for finite automata

An algorithm must be seen to be believed.

Donald Knuth.

30.1 Algorithmic problems for finite automata

In this section, we present algorithms for solving several natural problems about finite automata. We have already studied several algorithms for manipulating automata. For example, in Lecture 26 we examined algorithms for constructing deterministic finite automata for the intersection, union, and complementation operations. In Lecture 28, we learned about the subset construction, and the construction that transforms NFA with silent moves to NFA without silent moves. In this section, we describe some other interesting problems about automata and their solutions.

The emptiness problem. This problem is formulated as follows.

Design an algorithm that, given a finite automaton $\mathcal{A}$, checks if $\mathcal{A}$ accepts any string at all.

The problem is easy to solve by noticing the following. For the automaton $\mathcal{A} = (S, I, T, F)$, $\mathcal{A}$ accepts a string if and only if there exists a path from some initial state in I to some accepting state in F. Indeed, if $\mathcal{A}$ accepts a string w then $\mathcal{A}$ has an accepting run s_1, s_2, ..., s_{n+1} on w. The accepting run is simply a path from the initial state s_1 to the accepting state s_{n+1}. Conversely, suppose that there exists a path from some initial state to some accepting state:

$$s_1, s_2, \ldots, s_{m+1}$$

Let $\sigma_1, \ldots, \sigma_m$ be the labels of the edges $(s_1, s_2), \ldots, (s_m, s_{m+1})$. Then $\sigma_1 \ldots \sigma_m$ is a string accepted by $\mathcal{A}$. Thus, $\mathcal{A}$ accepts a string if and only if there exists a path from some initial state in I to a final state in F. We now describe an algorithm that solves the emptiness problem.

First, remove the edge labels in the transition diagram of the automaton $\mathcal{A}$. This produces a directed graph G (that depends on $\mathcal{A}$). Second, for every pair of states s, t where $s \in I$ and $t \in F$, check if there is a path from s to t. This can be done using the $PathExistence(G, s, t)$ algorithm from Lecture 7. Finally, if for some $s \in I$ and $t \in F$, there is a path from s to t, then $\mathcal{A}$ accepts at least one string. Otherwise, $\mathcal{A}$ accepts no string. This solves the emptiness problem for automata.

The equivalence problem. The formulation of the problem is this.

Design an algorithm that, given finite automata $\mathcal{A}$ and $\mathcal{B}$, decides whether these two automata are equivalent.

If we think of $\mathcal{A}$ and $\mathcal{B}$ as finite state descriptions of two programs, then this problem asks whether these two programs are equivalent. Recall that a state s in an automaton is *non-deterministic* if there is more than one transition from s labeled with the same input symbol. We describe one possible procedure, called $Equivalence(\mathcal{A}, \mathcal{B})$-algorithm, to solve this problem:

Algorithm 30.1 *Equivalence*$(\mathcal{A}, \mathcal{B})$ algorithm

---

(1) If $\mathcal{A}$ has nondeterministic states, then apply the subset construction to $\mathcal{A}$. Set $\mathcal{A}$ to be the newly constructed deterministic automaton.

(2) If $\mathcal{B}$ has nondeterministic states, then apply the subset construction to $\mathcal{B}$. Set $\mathcal{B}$ to be the newly constructed deterministic automaton.

(3) Construct two deterministic finite automata $\mathcal{A}^{(c)}$ and $\mathcal{B}^{(c)}$ that recognize the complements of $L(\mathcal{A})$ and $L(\mathcal{B})$, respectively.

(4) Construct two deterministic finite automata $\mathcal{A}_1$ and $\mathcal{B}_1$ that recognize the languages $L(A) \cap L(\mathcal{B}^{(c)})$ and $L(\mathcal{B}) \cap L(\mathcal{A}^{(c)})$, respectively.

(5) If $L(\mathcal{A}_1) \neq \emptyset$ then output *"$\mathcal{A}$ and $\mathcal{B}$ are not equivalent"*, and then stop.

(6) If $L(\mathcal{B}_1) \neq \emptyset$ then output *"$\mathcal{A}$ and $\mathcal{B}$ are not equivalent"*, and then stop. Otherwise, output *"$\mathcal{A}$ and $\mathcal{B}$ are equivalent"*.

Note that the automata $\mathcal{A}_1$ and $\mathcal{B}_1$ built at Step (4) of the algorithm recognize the languages $L(\mathcal{A}) \setminus L(\mathcal{B})$ and $L(\mathcal{B}) \setminus L(\mathcal{A})$, respectively. This algorithm is correct because $\mathcal{A}$ and $\mathcal{B}$ are equivalent if and only if $L(\mathcal{A}) \setminus$

$L(\mathcal{B}) = \emptyset$ and $L(\mathcal{B}) \setminus L(\mathcal{A}) = \emptyset$. This is exactly what the algorithm above detects.

The first two steps of the algorithm use subset construction. If an NFA has n states, then the subset construction produces a DFA with 2^n states. Therefore, it takes an exponential amount of time with respect to the numbers of states of the input automata $\mathcal{A}$ and $\mathcal{B}$ to execute the first two steps of the algorithm. If $\mathcal{A}$ and $\mathcal{B}$ are deterministic, then this algorithm is quite efficient because there is no need to run the subset construction.

The universality problem. This problem is formulated as follows.

Design an algorithm that, given a finite automaton $\mathcal{A}$, decides whether the automaton $\mathcal{A}$ recognizes Σ^*.

Notice that a language is Σ^* if and only if the complement of the language is empty. So, the universality problem has a straightforward solution. If $\mathcal{A}$ has nondeterministic states, we first apply the subset construction to $\mathcal{A}$. So, we can now assume that $\mathcal{A}$ is a deterministic finite automaton. We construct the automaton $\mathcal{A}^{(c)}$, which recognizes the complement of the language $L(\mathcal{A})$. Finally, we check if $L(\mathcal{A}^{(c)}) = \emptyset$. If $L(\mathcal{A}^{(c)}) = \emptyset$ then $\mathcal{A}$ recognizes Σ^*, and otherwise not.

As in the solution of the equivalence problem, our solution here involves the subset construction technique, which slows down the efficiency of the algorithm significantly.

The infinity problem. This problem is a bit more subtle and requires a better analysis of the state space of the automaton. The problem is the following.

Design an algorithm that, given a finite automaton $\mathcal{A}$, decides whether the automaton accepts infinitely many strings.

In order to solve this problem, we need to have a finer analysis of the transition diagrams of automata. We solve this problem in the next section, after we present a result that is interesting in its own right. The result is known as the pumping lemma.

30.2 The pumping lemma

The essence of the pumping lemma is that any sufficiently long run of an automaton must have repeated states. The reason for this is that the automaton has a fixed amount of memory, that is, a finite number of states. Hence, while making a long run, the automaton has to revisit one of the states that the automaton has already visited. Therefore, the segment of the run between any two repeated states can be performed any number of times without affecting the outcome of the run.

Lemma 30.1 (The Pumping lemma). *Let $\mathcal{A} = (S, I, T, F, \Sigma)$ be an NFA with exactly n states. Let $w = \sigma_1 \ldots \sigma_m$ be a string from Σ^* such that $m \geq n$. If $\mathcal{A}$ accepts the string w then there exists a nonempty substring $v = \sigma_i \ldots \sigma_{i+k-1}$ of the string w such that for all $t \geq 0$, the automaton $\mathcal{A}$ accepts the string*

$$\sigma_1 \ldots \sigma_{i-1} \underbrace{v \cdot v \cdot \ldots \cdot v}_{t \ \ times} \sigma_{i+k} \ldots \sigma_m.$$

Proof.
Since $\mathcal{A}$ accepts w, there exists a run

$$s_1, \ s_2, \ \ldots, \ s_m, \ s_{m+1}$$

of the automaton on the input $w = \sigma_1 \ldots \sigma_m$ such that $s_{m+1} \in F$. The key point here is that $m + 1 > n$, and hence, at least two states in the sequence $s_1, s_2, \ldots, s_m, s_{m+1}$ must be the same. Therefore, there must exist a state in this run that is repeated. Hence, there exist i and $i + k$ both less than or equal to $m+1$ such that $s_i = s_{i+k}$ and $k \neq 0$. Let $\bar{s}$ be the sequence of states:

$$s_i, \ \ldots, \ s_{i+k-1}.$$

Note that $s_i, \ldots, s_{i+k-1}, s_i$ is a run of the automaton on string $\sigma_i \ldots \sigma_{i+k-1}$ starting and ending at s_i. Recall that $\bar{s}^t$ denotes the concatenation of the sequence $\bar{s}$ with itself t times. We can now conclude that for any integer $t \geq 0$ the sequence

$$s_1, \ \ldots, \ s_{i-1}, \ s_i, \ \bar{s}^t, \ s_{i+k}, \ \ldots, \ s_{m+1}$$

is an accepting run of the automaton $\mathcal{A}$ on the input

$$\sigma_1 \ldots \sigma_{i-1} v^t \sigma_{i+k} \ldots \sigma_m,$$

where v is $\sigma_i \ldots \sigma_{i+k-1}$. $\qquad \square$

An interesting observation in this lemma is this. Consider the string $w = \sigma_1 \ldots \sigma_n$ from the lemma. We can write w as $w = u_1 v u_2$ such that $v \neq \lambda$ and $u_1 v^t u_2 \in L$ for all $t \in \mathbb{N}$. Hence, when $t > 1$, we have $|u_1 v^t u_2| > |w|$. When $t = 0$, we have $u_1 v^0 u_2 = u_1 u_2 \in L$ and $|u_1 u_2| < |w|$.

In the first case, we pumped the string w "up" and kept the pumped-up string in L. In the second case, we pumped the string w "down" and still kept the pumped-down string in L. In particular, if L is recognized by an n-state automaton then L must contain a string w such that $|w| < n$.

The Pumping lemma and its proof can be applied to obtain some interesting results. As an example, we provide an algorithm that detects whether an automaton $\mathcal{A}$ accepts infinitely many strings. We assume that $\mathcal{A}$ does not have silent transitions (by applying the construction from the previous lecture that removes all λ-transitions).

Now suppose that $L(\mathcal{A})$ is an infinite language. Then, there exists a string w accepted by $\mathcal{A}$ such the length of w is greater than the number of states in $\mathcal{A}$. From the proof of the Pumping lemma, we see that there exists a run $s_1, s_2, \ldots, s_m, s_{m+1}$ of $\mathcal{A}$ on w with the following properties:

(1) s_1 is an initial state and s_{m+1} is an accepting state.
(2) There exist i and $i + k$ with $k \neq 0$ such that $s_i, \ldots, s_{i+k}$ is a path and $s_i = s_{i+k}$. We call the state s_i a **looping state**.

Thus, if $L(\mathcal{A})$ is infinite then the automaton $\mathcal{A}$ has the following property. There exists a path from an initial state to a looping state and from the looping state to an accepting state.

Conversely, assume that there exists a path from an initial state to a looping state and from the looping state to an accepting state in $\mathcal{A}$. Then, $L(\mathcal{A})$ accepts infinitely many strings. Indeed, we take a string u_1 that labels the path from the initial state to the looping state, say q. Since q is a looping state, there must exist a nonempty string v that labels a path from q to itself. Next, take a string u_2 that labels a path from the looping state to an accepting state. Thus, we have that the strings $u_1 v^t u_2$ are all accepted by $\mathcal{A}$, where $t \geq 0$. Therefore, $L(\mathcal{A})$ is an infinite language. Based on this observation, the *Infinity*$(\mathcal{A})$ algorithm below solves the infinity problem.

Algorithm 30.2 *Infinity*$(\mathcal{A})$ algorithm

---

(1) Transform $\mathcal{A}$ into an NFA with no silent moves.
(2) Construct the set $X = \{q \mid q$ is a looping state $\}$.
(3) Check if there exists a path from an initial state s to a state $q \in X$ and a path from q to an accepting state. If there is such a path, then output "$L(\mathcal{A})$ *is infinite*". Otherwise, output "$L(\mathcal{A})$ *is finite*".

We note that Steps (2) and (3) of the algorithm can be implemented by using the $PathExistence(G, s, t)$ algorithm.

The pumping lemma can also be applied to prove that certain languages are not FA recognizable. A typical example is an elegant proof that the language

$$L = \{a^n b^n \mid n \in \mathbb{N}\}$$

over the alphabet $\{a, b\}$ is not FA recognizable. We already discussed this language in Lecture 26, and presented an informal explanation of why L is not FA recognizable. Here is now a formal proof of this fact.

Assume to the contrary that L is FA recognizable. Let $\mathcal{A}$ be a finite automaton with k states that recognizes L. Consider the string $u = a^k b^k$. This string satisfies the hypothesis of the pumping lemma (i.e. it is accepted by $\mathcal{A}$ and its length is greater than the number of states of $\mathcal{A}$). Therefore, by the Pumping lemma, we can split u into three pieces $u = xyz$, such that $|y| \geq 1$ and for all $t \geq 0$, the string $xy^t z$ is in L. There are three cases to consider.

Case 1. The string y contains a's only. Then, of course, the string $xyyz$ has more a's than b's. By the definition of L, the string $xyyz$ is *not* in L. However, by the pumping lemma $xyyz$ is in L. This is a contradiction.

Case 2. The string y contains b's only. As in *Case 1*, this also leads to a contradiction.

Case 3. The string y contains both a's and b's. Then, the string $xyyz$ must have some b's before a's. Such a string is not in L, by the definition of L. However, by the pumping lemma, the string $xyyz$ is in L. Again, we have a contradiction.

All these cases show that $\mathcal{A}$ cannot recognize L. Hence L is not FA recognizable.

30.3   The SAT problem and automata

We can convert the SAT problem (discussed in Lecture 24) to a problem about finite automata. Recall that the SAT problem asks if a given formula ϕ (of propositional logic) is satisfiable. The idea to solve this problem is the following. We construct a finite automaton $\mathcal{M}_\phi$ such that the formula ϕ is satisfiable if and only if the automaton $\mathcal{M}_\phi$ accepts at least one string.

We fix the first n propositions, and enumerate them in the following order:

$$p_1, \ p_2, \ p_3, \ \ldots, \ p_n.$$

Every truth assignment A of these propositions can now be identified with a binary string of length n over the alphabet $\{T, F\}$, where T is identified with **true** and F is identified with **false**. For example, when $n = 3$, all the truth assignments A of these propositions can be identified with the following binary strings:

$$TTT, \ TTF, \ TFT, \ TFF, \ FTT, \ FTF, \ FFT, \text{ and } FFF.$$

There are 2^n strings of length n and so there are 2^n truth assignments of the propositions

$$p_1, \ p_2, \ p_3, \ \ldots, \ p_n.$$

For example, the string $TT \ldots T$ is the assignment that assigns each p_i the value T, and the string $FTFT \ldots$ assigns T to every proposition p_{2i} and F to every proposition p_{2i+1}.

Let ϕ be a propositional logic formula such that the formula is built from propositions

$$p_1, \ p_2, \ p_3, \ \ldots, \ p_n.$$

We want to build a finite automaton $\mathcal{M}_\phi$ such that the automaton accepts exactly those truth assignments (and hence strings over the alphabet $\{T, F\}$) that evaluate the formula ϕ to **true**. In particular, if ϕ is not satisfiable then $\mathcal{M}_\phi$ accepts no string at all.

Our transformation from ϕ to $\mathcal{M}_\phi$ is by induction on the formula ϕ. Below we build $\mathcal{M}_\phi$, and at the same time prove that

$$L(\mathcal{M}_\phi) = \{A \mid A \text{ is a truth assignment that evaluates } \phi \text{ to } \textbf{true}\}.$$

To prove this equality we will be using the constructions for automata recognizing the union, intersection, and complementation operations.

Base case: Assume that ϕ is the proposition p_i. Then the assignment (string)

$$\sigma_1 \ldots \sigma_i \ldots \sigma_s \in \{T, F\}^\star$$

evaluates ϕ to **true** if and only if $\sigma_i = T$.

Clearly, we can construct a DFA $\mathcal{A}_\phi$ such that

$$L(\mathcal{A}_\phi) = \{A \mid A \text{ is a truth assignment that evaluates } \phi \text{ to } \textbf{true}\}.$$

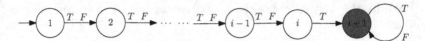

Fig. 30.1: The automaton accepting all assignments that evaluate the atomic formula $\phi = p_i$ to **true**.

The automaton is depicted in Figure 30.1. Clearly the automaton accepts all the assignments (strings) A that evaluate ϕ to **true**. The automaton is nondeterministic and has $i + 1$ states.

Inductive step: There are four cases to consider.

Case 1. Assume $\phi = (\phi_1 \vee \phi_2)$. By induction hypothesis, we have the following equalities:

$$L(\mathcal{M}_{\phi_1}) = \{A \mid A \text{ is a truth assignment that evaluates } \phi_1 \text{ to } \textbf{true}\},$$

and

$$L(\mathcal{M}_{\phi_2}) = \{A \mid A \text{ is a truth assignment that evaluates } \phi_2 \text{ to } \textbf{true}\}.$$

Set $\mathcal{M}_\phi$ to be the automaton that recognizes $L(\mathcal{M}_{\phi_1}) \cup L(\mathcal{M}_{\phi_2})$. This automaton then has the desired property:

$$L(\mathcal{M}_\phi) = \{A \mid A \text{ is a truth assignment that evaluates } \phi \text{ to } \textbf{true}\}.$$

Case 2. Assume $\phi = (\phi_1 \wedge \phi_2)$. By induction hypothesis, we have the following equalities:

$$L(\mathcal{M}_{\phi_1}) = \{A \mid A \text{ is a truth assignment that evaluates } \phi_1 \text{ to } \textbf{true}\},$$

and

$$L(\mathcal{M}_{\phi_2}) = \{A \mid A \text{ is a truth assignment that evaluates } \phi_2 \text{ to } \textbf{true}\}.$$

Set $\mathcal{M}_\phi$ to be the automaton that recognizes $L(\mathcal{M}_{\phi_1}) \cap L(\mathcal{M}_{\phi_2})$. This automaton is a desired automaton.

Case 3. The case for $\phi = (\phi_1 \rightarrow \phi_2)$ is similar to the cases above. The desired automaton must recognize the language

$$(\{T, F\}^\star \setminus L(\mathcal{M}_{\phi_1})) \cup L(\mathcal{M}_{\phi_2})$$

Case 4. Assume that $\phi = \neg\phi_1$. By induction hypothesis,

$$L(\mathcal{M}_{\phi_1}) = \{A \mid A \text{ is a truth assignment that evaluates } \phi_1 \text{ to } \textbf{true}\}.$$

Therefore, the automaton $\mathcal{M}_\phi$ that recognizes $\{T, F\}^* \setminus L(\mathcal{M}_{\phi_1})$ is the one desired.

The construction above proves the following theorem:

Theorem 30.1. *There exists an algorithm that, given a formula ϕ of the propositional logic, constructs a finite automaton $\mathcal{M}_\phi$ such that ϕ is satisfiable if and only if $L(\mathcal{M}_\phi) \neq \emptyset$.*

The transformation of formulas ϕ to the automata $\mathcal{M}_\phi$ tells us the following. The formula ϕ is satisfiable if and only if $L(\mathcal{M}_\phi) \neq \emptyset$. Thus, the SAT problem is reduced to the emptiness problem for automata. We discussed that the emptiness problem for automata can be solved efficiently. We also mentioned, in Lecture 24, that it is a major open question that asks to find an *efficient* algorithm to solve the SAT problem. Our reduction from SAT to the emptiness problem for automata may suggest that we have an efficient algorithm that solves the SAT problem. However, a hidden fact here is that the construction of $\mathcal{M}_\phi$ from ϕ can be quite expensive (that is, inefficient). The intuition here is the following. Assume that ϕ is a conjunction of m formulas $(\phi_1 \wedge \ldots \wedge \phi_m)$. Then $\mathcal{M}_\phi$ is the automaton that recognizes the language

$$L(\mathcal{M}_{\phi_1}) \cap \ldots \cap L(\mathcal{M}_{\phi_m}).$$

If each $\mathcal{M}_{\phi_i}$ has, say, m states then the automaton $\mathcal{M}_\phi$ has m^m states. This observation indicates that our reduction of the SAT problem to the emptiness problem for finite automata is quite inefficient, as expected.

30.4 Exercises

(1) Show that the language $\{a^n b^n c^n \mid n \in \mathbb{N}\}$ is not FA recognizable.
(2) Show that the language $\{ww \mid w \in \{a, b\}^*\}$ is not FA recognizable.
(3) Show that the language $\{w \mid w$ has an equal number of a's and b's $\}$ is not FA recognizable.
(4) Draw the transition diagram of the automaton $\mathcal{M}_\phi$ for each of the following propositional formulas:
$$p_1, \ p_2, \ (p_1 \wedge p_2), \text{ and } (p_2 \to (p_1 \wedge p_2)).$$
(5) Consider the algorithm that solves the emptiness problem as described in Section 30.1. Discuss its efficiency and if the algorithm can be improved.

Programming exercises

(1) Write a program that solves the emptiness problem.
(2) Implement the *Equivalence*($\mathcal{A}, \mathcal{B}$) algorithm.
(3) Write a program that solves the universality problem.
(4) Implement the *Infinity*($\mathcal{A}$) algorithm.

Lecture 31

Counting principles

In three words I can sum up everything
I've learned about life: It goes on.
Robert Frost.

31.1 Basic counting rules and principles

Once we have defined abstract objects, such as graphs, trees, or words, it is often the case that we would like to count the number of objects with specified properties. For example, let us call a string a *login* if its length is greater than 4 but less than 10, it starts with a letter (either lowercase or uppercase), and can contain letters, digits, and some symbols (namely _ or .). One would naturally like to count the number of logins possible to calculate the maximum number of users that can be supported. Counting concepts and techniques are used in analysis of algorithms, counting probabilities, and computer simulations. They also help us to reason about the abstract objects defined. In this lecture, we develop basic techniques and methods for counting.

We start with two obvious rules that are used in counting. The first rule is the **sum rule**. For this rule, recall that two sets A and B are *disjoint* if $A \cap B = \emptyset$. This rule states that if A_1, ..., A_n are pairwise disjoint sets with k_1, ..., k_n elements, respectively, then the union set $A_1 \cup \ldots \cup A_n$ has $k_1 + \ldots + k_n$ elements.

This rule is used in many applications of counting techniques. One can reformulate the rule as follows. Suppose we are given a task T, which can be completed by performing any one of the tasks T_1, ..., T_n. Each task T_i can be completed in k_i different ways, where $i = 1, \ldots, n$. Then there are

$k_1 + \ldots + k_n$ ways to perform the main task T.

Example 31.1. A secretary has one key, and she needs to label the key by either a letter or by a number strictly between 0 and 10. How many ways are there to label the key?

To solve the problem, we note that there are 26 ways to label the key with a letter, and 9 ways to label the key with a number. Thus, the key can be labeled in $26 + 9 = 35$ ways.

The second rule is the **product rule**. This rule states that if $A_1, \ldots, A_n$ are sets with $k_1, \ldots, k_n$ elements, respectively, then the set $A_1 \times \ldots \times A_n$, which is the Cartesian product of sets $A_1, \ldots A_n$, has $k_1 \cdot \ldots \cdot k_n$ elements.

It is easy to see that the rule is true (see Exercise 2). As the sum rule, the product rule has many interesting applications. In order to present some of these applications, we would like to reformulate the rule as follows.

Suppose we are given a task T of constructing some objects. These objects can be graphs, trees, words, functions, or tables. Suppose that we can break the task T into n subtasks $T_1, T_2, \ldots, T_n$ such that the following properties are true:

(1) to complete T all subtasks $T_1, \ldots, T_n$ must be completed in the given order,
(2) the subtask T_1 can be performed in k_1 ways, and
(3) for each i such that $1 < i \leq n$, after the subtasks $T_1, \ldots, T_{i-1}$ have been completed, the subtask T_i can be performed in k_i ways.

Then the task T can be executed in $k_1 \cdot \ldots \cdot k_n$ ways.

Example 31.2. Suppose you own 21 shirts, 10 pairs of pants, and 35 pairs of shoes. How many ways are there for you to get dressed in the morning?

The selection task consists of three subtasks. The first subtask is to select a shirt, the second is to select pants, and the third is to select shoes. We are, of course, assuming that you will indeed wear pants, a shirt, and shoes. There are 21 ways to select a shirt. After a shirt is selected, there are 10 ways to select pants. Finally, once you have picked your shirt and pants, there are 35 ways to select shoes. Thus, in total there are $21 \cdot 10 \cdot 35 = 7350$ different outfits.

Example 31.3. Let A and B be sets with n and m elements, respectively. How many functions are there from A to B?

Assume $A = \{a_1, \ldots, a_n\}$ and $B = \{b_1, \ldots, b_m\}$. Our task is to construct a function from A into B. For that we need to find an image for every $a \in A$. There are m possible images for a_1. Once the image of a_1 has been selected, there are m possible ways to select an image for a_2. Once the images of $a_1, \ldots, a_i$ have been selected there are m possible ways to select an image for a_{i+1}. Thus, by the product rule there are m^n ways to construct different functions. Hence, there are m^n functions from A to B.

Hence, there are 2^n functions from an n element set A into $\{0, 1\}$.

Example 31.4. How many bijective functions are there from A to A if A is an n element set?

As above, assume $A = \{a_1, \ldots, a_n\}$. Our task is to construct a bijective function from A to A. For that we need to find an image for every $a \in A$ such that no two images are the same. There are n possible images for a_1. Once an image of a_1 is selected there are $n - 1$ possible ways to select an image for a_2. Once the images of $a_1, \ldots, a_{i-1}$ are selected there are $n + 1 - i$ possible ways to select an image for a_i. Thus, by the product rule there are $n \cdot (n - 1) \cdot (n - 2) \cdots 2 \cdot 1$ ways to construct different bijective functions. Hence there are $n!$ bijective functions from A to A.

We now describe two counting principles. The first principle is known as the **pigeonhole principle**. The principle states the following. If more than n objects have been distributed into n boxes then one of the boxes must contain more than one object.

The proof of this principle is obvious. Indeed, if each box contains no more than one object then the number of objects must be less than or equal to n. However, we have more than n objects.

The principle can be reformulated as follows. Assume that we have distributed some objects into boxes such that each box contains no more than one object. Then the number of boxes is at least the number of objects.

Example 31.5. Let X be a set of positive integers with more than p elements, where $p > 0$. There exist two members of X that are congruent to each other modulo p.

To explain why this is true, let us assume that $X = \{n_1, \ldots, n_k\}$. We know that $k > p$. Consider p boxes labeled by $0, 1, \ldots, p - 1$. We put a number $x \in X$ into box i if the remainder of x when divided by p is i, that is x is congruent to i modulo p. Thus, we have distributed k numbers into

p boxes. By the pigeonhole principle, one of the boxes must contain two numbers from X. These two numbers must be congruent.

We can also recast the pigeonhole principle using the notion of injective function that we studied in Lecture 21. Indeed, let A be the set of objects (that we want to place into boxes) and B be the set of boxes. Then the pigeonhole principle states that if there exists an injection from the set A into the set B then A has less number of elements than the set B.

Example 31.6. Consider an arbitrary sequence of positive integers k_1, k_2, ..., k_{n+1} all less than or equal to $2n$, where n is a fixed positive integer. There are two numbers in this sequence such that one of them is a factor of the other.

To explain why this is true we use the pigeonhole principle. We write each number k_i as $2^{s_i} \cdot t_i$ such that t_i is an odd number and $s_i \geq 0$. To calculate t_i, one can divide k_i by 2, repeatedly, until the result is an odd number, and this gives us t_i. For instance, $24 = 2^3 \cdot 3$, and $23 = 2^0 \cdot 23$. Now consider the sequence $t_1, t_2, \ldots, t_{n+1}$. Each member t_i of this sequence is odd and is less than or equal to $2n$. There are exactly n odd numbers less than or equal to $2n$. By the pigeonhole principle, there must exist two distinct i and j such that $t_i = t_j$. Thus, we have $k_i = 2^{s_i} \cdot t_i$ and $k_j = 2^{s_j} \cdot t_j$ such that $t_i = t_j$. Clearly, one of these two numbers is a factor of the other.

Let X be a finite set of positive integers. Then the sum of X, written $sum(X)$, is the sum of all numbers in X. For example $sum(\{2, 5, 7\}) = 14$. We use this notion in the next example.

Example 31.7. Let A be a 10-element set of positive integers such that each member of A is less than or equal to 100. There are two disjoint and non-empty subsets X and Y of A such that $sum(X) = sum(Y)$.

Consider all subsets X of A. There are $2^{10} = 1024$ subsets of A. Each subset X of A has at most 10 elements. The largest possible sum, $sum(X)$, is therefore the sum of the set $\{91, 92, \ldots, 100\}$; this sum equals 955. Thus, the values of $sum(X)$ are among 0, 1, ..., 955. Therefore, we have more subsets of A than the values of their sums. By the pigeonhole principle, there are two distinct subsets X and Y of A such that $sum(X) = sum(Y)$. If X and Y are disjoint sets then we are done. Otherwise, both X and Y have a common element, say n. Then the sums of the sets $X \setminus \{n\}$ and $Y \setminus \{n\}$ are equal. Therefore, by removing elements common to both X and Y, we produce two disjoints subsets of A with equal sums.

The second principle is the **inclusion-exclusion principle**. The principle states that there are exactly $n + m - k$ elements in the union set $A \cup B$ if the set A has n elements, the set B has m elements, and k is the number of elements in $A \cap B$.

As for the sum and the product rules, we can restate the inclusion-exclusion principle as follows. Suppose we have two tasks T_1 and T_2, and there are n and m ways to complete each task, respectively. There may be common ways to complete the tasks. Let k be the number of such ways. Then the number of ways to do either of the tasks is $n + m - k$.

Example 31.8. How many license plate numbers are there that either start with letter B or end with 92?

To solve the example, we first need a definition of a license plate number. A plate number is a string of length 6 that contains letters or numbers. Thus, plate numbers are strings over the alphabet with $26 + 10 = 36$ symbols. By the product rule, there are exactly 36^5 plate numbers that start with a B. Similarly, there are 36^4 plate numbers that end with 92. There are 36^3 plate numbers that start with a B and end with 92. By the inclusion-exclusion principle, there are $36^5 + 36^4 - 36^3$ plate numbers that either start with a B or end with 92.

31.2 Permutations

We recall that two sequences are equal if they have the same length and the same elements in the same positions. Let A be a set with n elements. A **permutation** of A is any sequence $(m_1, \ldots, m_n)$ of elements of A in which no element is repeated. Note that the length of the permutation is equal to the number of elements of A. For example, for a two-element set A, say $A = \{x, y\}$, there are two permutations, which are (x, y) and (y, x). For a three-element set A, say $A = \{a, b, c\}$, all its permutations are (a, b, c), (a, c, b), (b, a, c), (b, c, a), (c, a, b), and (c, b, a).

Definition 31.1. An r-**permutation** of A, where $1 \leq r \leq n$, is any sequence $m_1, \ldots, m_r$ of elements of A of length r in which no element is repeated. If A has n elements, the we set:

$$P(n, r) = \text{the number of } r\text{-permutations of } A.$$

For instance, $P(n,1) = n$, and $P(n,2) = n \cdot (n-1)$. Note that the definition of $P(n,r)$ does not depend on the set A but rather it depends on n the number of elements of the set A. The product rule allows us to explicitly calculate $P(n,r)$ as proved in the next theorem.

Theorem 31.1. *For each $n \geq 0$ and r with $1 \leq r \leq n$ we have*
$$P(n,r) = n \cdot (n-1) \cdot \cdots \cdot (n-r+1).$$

Proof.
We use the product rule. Our task is to count the number of ways to build an r-permutation. To build an r-permutation $(m_1, \ldots, m_r)$, firstly, we need to choose the first item of it. There are n ways to choose the first item m_1. After the first item has been chosen, there are $n-1$ ways to choose m_2. Similarly, there are $n-2$ ways to choose the third item after the first two items have been chosen. Thus, there are $n-r+1$ ways to choose the last item m_r. By the product rule, the number of r permutations $P(n,r)$ equals $n \cdot (n-1) \cdots (n-r+1)$. $\qquad\square$

We note that $P(n,n)$, the number of permutations of an n element set, is equal to $n!$

Example 31.9. There are $36 \cdot 35 \cdot 34 \cdot 33 \cdot 32 \cdot 31$ ways to build a 6-permutation out of 36 elements.

We can write $P(n,r)$ as follows:
$$P(n,r) = \frac{n!}{(n-r)!}.$$
Indeed,
$$P(n,r) = P(n,r) \cdot \frac{(n-r)!}{(n-r)!} = \frac{(n-r)! \cdot P(n,r)}{(n-r)!} = \frac{n!}{(n-r)!}.$$
We would like to study some interesting properties of permutations. Henceforth, we fix an n-element set A. Without loss of generality we assume that $A = \{1, \ldots, n\}$. We single out the set of all permutations of A in the next definition:

Definition 31.2. The set of all permutations of $A = \{1, \ldots, n\}$ is denoted by S_n.

Given a permutation $P = (m_1, \ldots, m_n)$ of $A = \{1, \ldots, n\}$ we can also represent it as follows:

$$P = \begin{pmatrix} 1 & 2 & \ldots & n \\ m_1 & m_2 & \ldots & m_n \end{pmatrix}$$

We call this a **standard representation** of the permutation. We can identify this permutation with the function $P : A \to A$ (also denoted by P) for which we have $P(i) = m_i$ where $i = 1, \ldots, n$. This function is a bijective function since all the entries in $(m_1, \ldots, m_n)$ are distinct. The opposite is also true. Namely, every bijection $Q : A \to A$ can be identified with the permutation $(k_1, \ldots, k_n)$ such that $Q(i) = k_i$ for $i = 1, \ldots, n$.

Given two permutations

$$P = \begin{pmatrix} 1 & 2 & \ldots & n \\ m_1 & m_2 & \ldots & m_n \end{pmatrix} \quad \text{and} \quad Q = \begin{pmatrix} 1 & 2 & \ldots & n \\ k_1 & k_2 & \ldots & k_n \end{pmatrix}$$

we can form the **product**, denoted by $P \circ Q$, of P and Q as follows. The value of $P \circ Q$ on i is computed by first applying Q to i, and then applying P to $Q(i)$ for $i = 1, \ldots, n$. Thus,

$$P \circ Q = \begin{pmatrix} 1 & 2 & \ldots & n \\ m_{k_1} & m_{k_2} & \ldots & m_{k_n} \end{pmatrix}$$

For example, for

$$P = \begin{pmatrix} 1\,2\,3\,4\,5 \\ 2\,4\,5\,3\,1 \end{pmatrix} \quad \text{and} \quad Q = \begin{pmatrix} 1\,2\,3\,4\,5 \\ 3\,5\,1\,2\,4 \end{pmatrix}$$

we have

$$P \circ Q = \begin{pmatrix} 1\,2\,3\,4\,5 \\ 5\,1\,2\,4\,3 \end{pmatrix}$$

If one thinks of P and Q as functions from A to A then $P \circ Q$ is simply the composition of the functions P and Q. Since composition of bijective functions is a bijection (as we know from Lecture 21), the product of permutations is again a permutation.

The identity permutation, denoted by I_A, is

$$I_A = \begin{pmatrix} 1\,2\,\ldots\,n \\ 1\,2\,\ldots\,n \end{pmatrix}$$

It is easy to check that $P \circ I_A = P$ and $I_A \circ P = P$ for all permutations P.

31.3   Exercises

(1) Solve the following problems using the sum and product rules:

   (a) There are 8 mathematics graduates and 22 computer science graduates to be interviewed by a company. How many ways are there
for the company to hire a graduate?
   (b) A multiple choice test contains 8 questions. The first three questions each have 3 possible answers, the next 3 questions have four
possible answers, and the last two have 2 possible answers. How
many ways can a student complete the test if the student answers
all the questions? How many ways are there if the student might
leave some questions unanswered?
   (c) How many strings are there of length k over an n-letter alphabet?
   (d) How many binary strings of length 10 are there if the string starts
with either 101 or 010 and ends with either 11 or 00?
   (e) A **palindrome** is a string w such that when we write w in reverse, the resulting string is again w. For example, 101, 1001, and
10100101 are palindromes. How many binary strings of length n
are palindromes?
   (f) Consider a truth table built from propositions p_1, ..., p_n. Each
entry of the table has T or F written in it. Prove that there are
2^{2^n} truth tables.

(2) Prove, using the induction principle, the correctness of the product
rule.
(3) Consider a group of people. How many people does the group need
to have to guarantee that at least two from the group are born in the
same month?
(4) Consider the sequence $t+1, t+2, \ldots, t+p$, where t is an integer. Show
that n is a factor for at least one of these integers if $p > n$.
(5) Let A, B, C, D, E, F be the grades given to a class of students. What
is the minimal number of students the class must have to guarantee
that at least four of the students get the same mark?
(6) A professional soccer team has 5 weeks to prepare for a final cup game.
The team trains at least once each day. The coach of the team decides
not to have more than 11 training sessions in any 7-day period. Prove
that there exists a sequence of successive days during which the team
has exactly 14 training sessions.
(7) Prove that in a group of $n \geq 2$ people there are two who know the same

number of people in the group. In the exercise we assume that the "to know relationship" is symmetric.

(8) Write down inclusion-exclusion principle for three and four sets.
(9) Prove that the composition of two permutations is a permutation.
(10) Prove that given a permutation P of set A, there exists a k such that P^k is the identity permutation I_A (P^k is the composition of P with itself k times).

Programming exercises

(1) Implement a function that takes a set A of objects and an integer r as input, and prints all r-permutations of A.
(2) Design a Permutation class for permutations in their standard representations.
(3) Implement a function that takes two Permutation objects, and outputs their composition.

Lecture 32

Permutations and combinations

Every possible permutation is still in play.

Michael Cullen.

32.1 Cycles and permutations

Our goal is to express a given permutation P on a set A as a composition of simpler permutations. For this we need to define the concept of a simple permutation. For the next definition, we think of permutations P of A as bijective functions, as explained in Lecture 31.

Definition 32.1. A permutation P of A is called a **cycle** if there are numbers $i_1, i_2, \ldots, i_k$ in A such that $P(i_1) = i_2$, $P(i_2) = i_3$, ..., $P(i_{k-1}) = i_k$ and $P(i_k) = i_1$, and $P(i) = i$ for all other elements $i \in A$. We represent the cycle as $[i_1, \ldots, i_k]$. The number k is the **length of the cycle**.

Here is an example of a cycle:

$$\begin{pmatrix} 1 \ 2 \ 3 \ 4 \ 5 \ 6 \ 7 \ 8 \\ 2 \ 4 \ 3 \ 6 \ 5 \ 8 \ 7 \ 1 \end{pmatrix}$$

Thus, this cycle is such that $P(1) = 2$ $P(2) = 4$, $P(4) = 6$, $P(6) = 8$ and $P(8) = 1$. Clearly, the length of the cycle is 5.

Let C be a cycle $[i_1, \ldots, i_k]$. We sometimes refer to the the elements i_1, ..., i_k as the element in the cycle. If there is no confusion, we also might identify the cycle C with the set $\{i_1, \ldots, i_k\}$. Note that the cycle C, as follows from the definition, does not move elements k outside of the cycle, that is, $C(j) = j$ for all $j \notin \{i_1, \ldots, i_k\}$.

Other examples of cycles are

$$\begin{pmatrix} 1 \ 2 \ .... \ i \ .... \ j \ .... \ n \\ 1 \ 2 \ .... \ j \ .... \ i \ .... \ n \end{pmatrix}$$

These are cycles of length 2 and are called **transpositions**. We denote this transposition by $[i, j]$. Thus, the transposition $[i, j]$ is the permutation P such that $P(i) = j$, $P(j) = i$, and $P(k) = k$ for all other elements $k \in A$.

To motivate the next theorem we analyze the following permutation

$$P = \begin{pmatrix} 1 & 2 & 3 & 4 & 5 & 6 & 7 & 8 & 9 & 10 \\ 3 & 4 & 8 & 10 & 7 & 5 & 6 & 1 & 2 & 9 \end{pmatrix}$$

For P we have the following:

  (a) $P(1) = 3$, $P(3) = 8$, $P(8) = 1$, and
  (b) $P(2) = 4$, $P(4) = 10$, $P(10) = 9$, $P(9) = 2$, and
  (c) $P(5) = 7$, $P(7) = 6$, and $P(6) = 5$.

Thus, P can be *decomposed* into three cycles. One cycle is of length 4 and two are of length 3. These cycles are $[1, 3, 8]$, $[2, 4, 10, 9]$, and $[5, 7, 6]$. We observe that P can be written as a composition of cycles in the following way:

$$P = [1, 3, 8] \circ [2, 4, 10, 9] \circ [5, 7, 6].$$

We generalize this observation in the next theorem:

Theorem 32.1. *Every permutation is a composition of cycles.*

Proof.
Consider a permutation P:

$$P = \begin{pmatrix} 1 & 2 & .... & n \\ m_1 & m_2 & .... & m_n \end{pmatrix}$$

Let us start forming the sequence

$$i_1 = 1, \ i_2 = P(i_1), \ i_3 = P(i_2), \ i_4 = P(i_3), ...$$

Consider the first stage k at which $P(i_k)$ has already appeared in the sequence. It must be the case that $P(i_k) = i_1$ because P is a permutation. Thus we have the cycle

$$C_1 = [i_1, ..., i_k].$$

If for all $j \notin C_1$, we have $P(j) = j$, then we have decomposed P into one cycle. Otherwise, we take the least element j_1 not in the cycle C_1 such that $P(j_1) \neq j_1$ and start forming the sequence

$$j_1, j_2 = P(j_1), \ j_3 = P(j_2), \ j_4 = P(j_3), \ldots.$$

Consider the first stage at which $P(j_t)$ has already appeared in the sequence. It must be the case that $P(j_t) = j_1$ because P is a permutation. Thus we have the cycle

$$C_2 = [j_1, \ldots, j_t].$$

The cycles C_1 and C_2 have no elements in common because P is a permutation. In other words, we have

$$\{i_1, \ldots, i_k\} \cap \{j_1, \ldots, j_t\} = \emptyset.$$

If for all $j \notin C_1 \cup C_2$, we have $P(j) = j$, then we have decomposed P into two cycles. Otherwise, we select the least element l_1 not in $C_1 \cup C_2$ such that $P(l_1) \neq l_1$ and form a new cycle starting from l_1 as above. We continue this process and produce cycles $C_1, C_2, \ldots, C_m$ such that every element j of A is either in one of these cycles or $P(j) = j$. Now we see that $P = C_1 \circ C_2 \circ \ldots \circ C_m$. $\qquad \square$

This theorem can be strengthened to show that every permutation is a composition of transpositions (see Exercise 3).

32.2   Combinations

In Lecture 31, we computed the number of r-permutations of an n-element set. In this section we compute the number of r-element subsets of an n-element set and prove some interesting facts about counting such subsets.

Definition 32.2. An **r-combination** of a set A with n elements is an r-element subset of A.

Thus, for $A = \{0, 1\}$ there is exactly one 0-combination (the empty set $\emptyset$), two 1-combinations (the sets $\{0\}$ and $\{1\}$), and one 2-combination (the set A). For a 3-element set $\{a, b, c\}$ there is exactly one 0-combination (the empty set $\emptyset$), three 1-combinations (the sets $\{a\}$, $\{b\}$, and $\{c\}$), three 2-combinations (the sets $\{a, b\}$, $\{a, c\}$, and $\{b, c\}$), and one 3-combination (the set A). As for r-permutations, we single out the following notation:

Definition 32.3. For a set A with n elements and $0 \leq r \leq n$, we set
$$C(n,r) = \text{the number of } r\text{-combinations of } A.$$

Thus, $C(n,r)$ represents the number of ways that r things can be chosen from a set of n things. Typically, $C(n,r)$ is read as "n choose r". For example, $C(n,0) = 1$ since there is exactly one subset of A with no elements. Similarly, $C(n,n) = 1$. Also, $C(n,1) = n$ since the number of 1-element subsets of A is n. We point out that the number $C(n,r)$ does *not* depend on the set A but rather depends on the number of elements of A.

Since any r-element subset X of A uniquely determines its complement $A \setminus X$ (which is an $(n-r)$-element subset of A), we clearly have the identity:
$$C(n,r) = C(n, n - r).$$

In the previous lecture, we gave an explicit formula for computing $P(n,r)$. It turns out that there is also an explicit formula for computing $C(n,r)$ given by the next theorem.

Theorem 32.2. *For each $n \geq 0$ and r with $0 \leq r \leq n$, we have the following:*
$$C(n,r) = \frac{n!}{r! \cdot (n-r)!}.$$

Proof.
The proof can be carried out by recasting the definition of r-permutation through the use of the product rule. The task of constructing an r-permutation can be viewed as follows. To construct an r-permutation we first need to select an r-combination, and then order the combination. Thus, we can apply the product rule as follows. There are $C(n,r)$ number of ways to select an r-combination. Once an r-combination is selected, there are $r!$ ways to order it. Therefore we have the equality $P(n,r) = C(n,r) \cdot r!$. Now we perform an easy calculation using the formula $P(n,r) = \frac{n!}{(n-r)!}$ obtained in the previous lecture:
$$C(n,r) = \frac{P(n,r)}{r!} = \frac{n!}{r! \cdot (n-r)!}.$$
We have the desired identity. $\qquad\square$

For example,

$$C(6,3) = \frac{6!}{3!(6-3)!} = \frac{1 \cdot 2 \cdot 3 \cdot 4 \cdot 5 \cdot 6}{1 \cdot 2 \cdot 3 \cdot 1 \cdot 2 \cdot 3} = 20$$

Example 32.1. Recall that an undirected graph is *complete* if there is an edge between every two vertices of the graph. Assume $\mathcal{G} = (V, E)$ is a complete graph with n vertices. Every edge of $\mathcal{G}$ is simply a two-element subset of V. Therefore there are $C(n, 2)$ number of edges. The exact number of edges is thus:

$$C(n,2) = \frac{n!}{2!(n-2)!} = \frac{n \cdot (n-1)}{2}.$$

One of the interesting properties of $C(n, r)$ is known as **Pascal's identity**. The identity establishes a relationship between r-combinations of n and $n + 1$ element sets.

Theorem 32.3. *For $n \geq 0$ and r with $0 \leq r \leq n$, the following Pascal's identity holds:*

$$C(n+1, r) = C(n, r-1) + C(n, r).$$

Proof.
Take a set A with $n + 1$ elements. We want to compute the number of all r-element subsets of A. Fix an element $b \in A$ and consider the set $X = A \setminus \{b\}$. The set X has n elements. Every r-element subset S of A either contains or does not contain b. Therefore it suffices to count the number of r-element subsets S of A that contain b, and those that do not contain b.

Suppose that $b \in S$. Then $S = \{b\} \cup S_1$, where S_1 is a $(r-1)$-element subset of X. The number of $(r-1)$-element subsets of X equals $C(n, r-1)$. Therefore, there are exactly $C(n, r-1)$ subset of A that contain b.

Suppose that $b \notin S$. Then S is a subset of X. Again, there are exactly $C(n, r)$ subsets of X by the definition of $C(n, r)$. Therefore, there are $C(n, r)$ subsets of A that do not contain b. Consequently, we have the Pascal's identity $C(n+1, r) = C(n, r-1) + C(n, r)$. $\square$

The numbers $C(n, r)$ are also called **binomial coefficients**. Binomial coefficients are used when one expands expressions of the type $(x + y)^n$, where x, y are variables. For example:

$$(x+y)^2 = x^2 + 2xy + y^2 \quad \text{and} \quad (x+y)^3 = x^3 + 3x^2y + 3xy^2 + y^3.$$

When the expression $(x+y)^n$ is expanded one obtains the terms of the form $x^{n-j}y^j$. For instance, when we expand the expression $(x+y)^4$ the terms we get are

$$x^4, \ x^3y, \ x^2y^2, \ xy^3, \text{ and } y^4.$$

The task is to calculate the coefficients in front of these terms. The theorem below tells us how we can calculate these coefficients.

Theorem 32.4. *Let n be a positive integer and x, y be integer variables. Then*

$$(x+y)^n = \sum_{j=0}^{n} C(n,j)x^{n-j}y^j.$$

Proof.
Consider the expression $(x+y)^n$. By definition, this expression is equal to

$$(x+y) \cdot (x+y) \cdot \ldots \cdot (x+y),$$

where the term $(x + y)$ appears exactly n times. When we expand this expression we have the terms of the form $x^{n-j}y^j$. In this expansion, each such term $x^{n-j}y^j$ is obtained by selecting x from $n - j$ terms of the form $(x+y)$ and by selecting y in the remaining j terms. Thus, by definition of $C(n,j)$, the number of terms of the type $x^{n-j}y^j$ must be equal to $C(n,j)$. Therefore the coefficient in front of $x^{n-j}y^j$ is $C(n,j)$. $\square$

32.3   Exercises

(1) Prove that for all permutations P, Q, and R it is always the case that
$$(P \circ Q) \circ R = P \circ (Q \circ R).$$

(2) Prove that for every permutation P there is a permutation Q such that $P \circ Q = Q \circ P = I_A$.

(3) Prove that every permutation P can be written as a composition of transpositions (that is 2-cycles). (Hint: by Theorem 32.1 it suffices to show that every cycle is a composition of transpositions).

(4) Prove that $C(n,k) = nC(n-1, k-1)/k$ for $k > 0$.

(5) Prove that $2^n = \Sigma_{k=0}^n C(n,k)$.

(6) Expand the expressions $(x+y)^5$ and $(a+b+c)^5$.

(7) Prove that if a permutation P is written as a composition of two cycles $P = C_1 \circ C_2$ and $C_1 \cap C_2 = \emptyset$ then P can also be decomposed as $C_2 \circ C_1$.

(8) Give examples of permutations P and Q such that $P \circ Q \neq Q \circ P$.

(9) The aim of this exercise is to illustrate how counting techniques can be applied to card games.

In a deck of cards there are four suits: *clubs, diamonds, hearts, and spades.* Each suit consists of thirteen cards whose values are A (ace), 2, 3, 4, 5, 6, 7, 8, 9, 10, J (jack), Q (queen), and K (king). For any given value there are four cards, one from each suit. Hence, there are exactly 52 cards.

A **poker hand** is a set of 5 cards (out of 52 cards). A **straight** is a poker hand in which the card values can be arranged to form a consecutive sequence. For example, $8, 9, 10, J, Q$ is a straight. The ace A can be at the bottom of a straight (e.g. $A, 2, 3, 4, 5$) or at the top of a straight (e.g. $10, J, Q, K, A$). Now we define several types of poker hands:

Royal flush: $10, J, Q, K, A$ all of the same suit.

Straight flush: Straight all in the same suit which is not Royal flush.

Four of a kind: Four cards in the hand have the same value.

Full house: Three cards have one value and the other two cards have another value.

Flush: All cards are of the same suit, but not a straight or Royal flush.

Three of a kind: Three cards are of one value, a fourth card is another value, and a fifth card is of a third value.

Two pairs: Two cards are of one value, another two cards are of another value, and the remaining card is of a third value.

One pair: Two cards are of one value, and three other cards have three different values.

Nothing: None of the above.

How many poker hands are there? Count the poker hands of the following types:

(a) Full house,

(b) Two pairs,

(c) Straight,

(d) Four of a kind,

(e) Flush,

(f) Three of a kind, and

(g) One pair.

Programming exercises

(1) Implement a method that takes an instance P of the Permutation class (as defined in a programming exercise in Lecture 31) as input, and outputs the cycles $C_1, \ldots, C_m$ such that $P = C_1 \circ C_2 \circ \ldots \circ C_m$.

(2) Implement a recursive algorithm to compute $C(n, r)$ (hint: use Pascal's Identity).

Lecture 33

Basics of probability

When it is not in our power to determine what is true,
we ought to follow what is most probable.
René Descartes.

33.1 Elements of probability

Probability theory is a branch of mathematics that studies uncertainty. The mathematical roots of the theory go back to the work of Pierre de Fermat and Blaise Pascal in the seventeenth century. Modern probability theory was founded by Andrey Kolmogorov, one of the most famous Soviet mathematicians of the twentieth century.

Uncertainty and randomness occur in many systems that we design. Therefore, uncertainty should be taken into account when we design systems, such as communication systems, control systems, integrated circuits, and so on. For instance, when we design communication systems we need to consider noise effects, and when we design web servers we need to consider unpredictable traffic fluctuations. The goal of this lecture is to introduce some of the key concepts of probability theory with the help of many examples.

Suppose we have an experiment that yields various outcomes. For example, the experiment can be throwing a die with six possible outcomes which are 1, 2, 3, 4, 5, and 6. For a given experiment, the set of all possible outcomes is called the **sample space**. We denote sample spaces of experiments by Ω. The elements of the sample space Ω are called the **outcomes** of the experiment. Thus, in the case above, $\Omega = \{1, \ldots, 6\}$. Another experiment can be selecting a letter from the English alphabet. Each outcome of this experiment is a letter. The sample space Ω in this experiment consists of all the letters in the English alphabet. The next example of an experiment is when a pair of six sided dice is rolled. Each outcome of this experiment is a pair (i, j) of numbers where both i and j are 1, 2, 3, 4, 5 or 6. Thus, the sample space Ω of this experiment consists of 36 outcomes.

In a given experiment, we might be interested in certain events occurring. For instance, we might be interested in rolling an odd number on a die. An **event** is thus a subset of the sample space Ω. We denote events by letters such as E possibly with indices. If x is an outcome of the experiment, $E \subseteq \Omega$ is an event, and $x \in E$ then we say that event E has occurred. Typically, given two events E_1 and E_2 in a sample space Ω, we would like to know the probabilities of the following events occurring: E_1 or E_2, E_1 and E_2, and not E_1. To reason about probabilities we give the following main definition of this lecture.

Definition 33.1. The **probability on** Ω is a function P that assigns a value $P(E)$ for every subset of Ω such that the following conditions are satisfied:

(1) The probability that the event E occurs is between 0 and 1, that is
   $0 \le P(E) \le 1$.
(2) For disjoint events E_1 and E_2 (that is $E_1 \cap E_2 = \emptyset$) we have $P(E_1 \cup E_2) = P(E_1) + P(E_2)$.
(3) $P(\Omega) = 1$.

We call the pair (Ω, P) a **probability space**.

The idea in this definition is that $P(E)$ measures the chance of the event E occurring in the experiment. For instance, $P(E) = 0$ indicates that there is no chance that the outcome of the experiment will be in E. If $P(E) > 0$ then this means that there is some chance that event E occurs. The second part of the definition of P is known as the *additivity property*. The additivity property reflects the way we think probabilities work in real life. For example, in throwing a die the probability that 2 or 4 occurs is equal to the probability that 2 occurs in the outcome plus the probability that 4 occurs in the outcome. Every outcome of the experiment is in Ω, therefore the probability that Ω occurs is 1. This is stipulated in the last part of the definition. Finally, we note that there is no restriction on the sample space Ω to be finite. If Ω is infinite, then one needs to be a bit more careful and replace the second part of the definition with an infinitary formula. In this lecture however, we *always* assume that the sample space Ω is a finite set.

The following theorem collects several simple properties of probability spaces that can be derived from the definition. In the theorem, $|X|$ denotes the number of elements of set X.

Theorem 33.1. *Let (Ω, P) be a probability space. Let $E = \{e_1, \ldots, e_k\}$ be an event, that is $E \subseteq \Omega$. Then the following properties of P are true:*

(1) $P(E) = P(\{e_1, \ldots, e_k\}) = \sum_{e \in E} P(\{e\})$.
(2) $P(\Omega \setminus E) = 1 - P(E)$.
(3) If all the outcomes of Ω are equally likely to occur then

$$P(E) = \frac{|E|}{|\Omega|}.$$

Proof.
The first part follows from the additivity property of P. Indeed, $E = \{e_1\} \cup \ldots \cup \{e_k\}$. Clearly, each of these sets is disjoint from the others. Therefore, using the additivity property of P (and induction on k) we get:

$$P(E) = P(\{e_1\}) + \ldots + P(\{e_k\}) = \sum_{e \in E} P(\{e\}).$$

The second part again follows from the additivity property of P:

$1 = P(\Omega) = P(E \cup (\Omega \setminus E)) = P(E) + P(\Omega \setminus E)$. Hence, $P(\Omega \setminus E) = 1 - P(E)$.

For the last part we note the following. The hypothesis states that all the outcomes of Ω occur equally likely. This means that $P(\{e\}) = P(\{e'\})$ for all outcomes $e, e' \in \Omega$. Hence, by Part (1) we have

$$1 = P(\Omega) = \sum_{e \in \Omega} P(\{e\}). \quad \text{Hence } P(\{e\}) = \frac{1}{|\Omega|}.$$

Therefore,

$$P(E) = \sum_{e \in E} P(\{e\}) = \sum_{e \in E} \frac{1}{|\Omega|} = \frac{|E|}{|\Omega|}.$$

$\square$

The first part of the theorem states that knowing the probability of every outcome allows us to compute the probability of any given event E. The second part of the theorem gives us an explicit formula for computing the probability of event E *not* occurring. The last part of the theorem considers experiments in which outcomes are equally likely to occur. Throwing a die or tossing a coin are examples. For such experiments, computing the probability of E boils down to computing the number of total outcomes of the experiment and the number of outcomes in E. In the examples below all the outcomes are equally likely to occur.

Example 33.1. A class of students has 5 students of age seventeen, 12 students of age eighteen, and 8 students of age nineteen. What is the probability that a student chosen from the class is at age seventeen or nineteen?

To solve the problem note that there are $25 = 5 + 12 + 8$ students. So, the sample space consists of 25 outcomes. Thirteen of these are students at age seventeen or nineteen. Thus, probability of a chosen student to be at age seventeen or nineteen is $\frac{13}{25}$.

Example 33.2. Consider the experiment of throwing a die. The outcomes of the experiment are 1, 2, ..., 6. What is the probability that the outcome k of the experiment is even? What is the probability that the outcome k of the experiment is either 1 or 5?

To solve this problem note that the sample space Ω is $\{1, 2, 3, 4, 5, 6\}$. Therefore, for each $k \in \Omega$, we have $P(\{k\}) = \frac{1}{6}$. The event $E = \{2, 4, 6\}$ corresponds to even outcomes. The event $F = \{1, 5\}$ corresponds to the outcomes 1 or 5. Therefore $P(E) = \frac{3}{6} = \frac{1}{2}$ and $P(F) = \frac{2}{6} = \frac{1}{3}$.

The next theorem explores additional properties of probability spaces:

Theorem 33.2. *Let* (Ω, P) *be a probability space. Then the following properties of* P *are true:*

(1) $P(\emptyset) = 0$.
(2) For all events E *and* F *we have* $P(E \cup F) = P(E) + P(F) - P(E \cap F)$.
(3) For pairwise disjoint events E_1, ..., E_k, *we have:*

$$P(E_1 \cup \ldots \cup E_k) = P(E_1) + \ldots + P(E_k).$$

Proof.
The first part follows from the fact that $P(\emptyset) = P(\emptyset \cup \emptyset) = P(\emptyset) + P(\emptyset) = 2P(\emptyset)$. This happens only when $P(\emptyset) = 0$. For the second part, one notes the following obvious equalities:

$$E \cup F = E \cup (F \setminus E) \quad \text{and} \quad F = (F \setminus E) \cup (F \cap E).$$

The unions on the right sides of the above equalities are unions of disjoint sets. By the additivity property of P we have:

$$P(E \cup F) = P(E) + P(F \setminus E) \quad \text{and} \quad P(F) = P(F \setminus E) + P(F \cap E).$$

Putting these two together we have $P(E \cup F) = P(E) + P(F) - P(E \cap F)$. Part (3) can be proved by induction. The base of the induction $k = 2$ is the additivity property of P. $\qquad\square$

Note that part (2) of the theorem can be viewed as recasting the inclusion-exclusion principle for probability spaces (See Section 31.1).

Example 33.3. Fix $n \geq 1$. Suppose we flip a coin n times where the tail (T) and head (H) outcomes are equally likely to occur. So, the set of outcomes Ω consists of all n-tuples over $\{T, H\}$. It is reasonable to assume that all the outcomes in Ω are equally likely to occur.

(1) What is the probability of each outcome?
(2) What is the probability that that outcome has exactly r heads?

Clearly, Ω has 2^n outcomes. Since all the outcomes are equally likely to occur, the probability of each outcome is $\frac{1}{2^n}$. Consider the event E defined as follows:

$$E = \{(x_1, \ldots, x_n) \in \Omega \mid \text{the outcome } (x_1, \ldots, x_n) \text{ has exactly } r \text{ heads}\}.$$

A simple observation is that the number of outcomes with exactly r heads equals the number of r-element subsets of an n-element set. Therefore, from Lecture 32, there are exactly $C(n, r)$ outcomes with exactly r heads. Hence,

$$P(E) = C(n, r)/2^n.$$

33.2 Conditional probability

Suppose we have an experiment with sample space Ω, and that an event F has occurred. Given that F occurred we want to find the probability that a specified event E has also happened. The simple idea here is to treat

F as the sample space since we know that F has occurred. Therefore the probability of E, knowing that F has already occurred, should be computed relative to F rather than relative to the whole sample space Ω. Formally, we give the following definition:

Definition 33.2. Let (Ω, P) be a probability space. **The conditional probability of E given F** with $P(F) > 0$, denoted by $P(E \mid F)$, is defined as follows:

$$P(E \mid F) = \frac{P(E \cap F)}{P(F)}.$$

We refer to this formula as the *conditional probability formula*.

Example 33.4. Suppose we flip a coin 4 times where the tail and head outcomes are equally likely to occur. We would like to know the probability that an even number of tails appear given that the second flip is tail.

There are 16 outcomes of the experiment. Let F be the event that the second flip is tail. There are exactly 8 outcomes in F. Hence $P(F) = 0.5$. Also, there are 8 outcomes of the experiment with an even number of tails. Let us denote this event by E. Among these 8 outcomes 4 outcomes belong to the event F. These are $TTTT, TTHH, HTTH$, and $HTHT$. Therefore $P(E \cap F) = 0.25$. Thus,

$$P(E \mid F) = \frac{P(E \cap F)}{P(F)} = \frac{0.25}{0.5} = \frac{1}{2}.$$

Here is another interesting example.

Example 33.5. Robert and Indira have two children. One child is a girl. What is the probability that the other child is a girl?

The first guess is that the probability is 0.5 since the second child being a girl does not depend on the first child's gender. However, more detailed analysis shows that the guess is incorrect. Indeed, the sample space Ω in this example is the following: $\{BB, GB, BG, GG\}$, where B indicates boy and G indicates girl. All of these outcomes are equally likely to occur. Hence, $P(BB) = P(GB) = P(BG) = P(GG) = 0.25$. Let F be the event $\{GG, BG, GB\}$, and E be the event $\{GG\}$. We know that the event F has occurred by the assumption. Therefore, to calculate the probability of the other child being a girl, we use the conditional probability formula:

$$P(E \mid F) = \frac{P(E \cap F)}{P(F)} = \frac{1/4}{3/4} = \frac{1}{3}.$$

The next theorem lists several properties of conditional probability.

Theorem 33.3. *Let (Ω, P) be a probability space and F be an event such that $P(F) > 0$. Then each of the following is true:*

(1) $P(F \mid F) = 1$.

(2) For all events E, we have $0 \leq P(E \mid F) \leq 1$.

(3) If $E_1 \cap E_2 = \emptyset$ with $E_1 \subseteq F$ and $E_2 \subseteq F$, then

$$P(E_1 \cup E_2 \mid F) = P(E_1 \mid F) + P(E_2 \mid F).$$

(4) For events E such that $E \subseteq F$ we have $P(E \mid F) \geq P(E)$.

Proof.
The first two properties follows easily from the conditional probability formula. For (3), for pairwise disjoint events E_1 and E_2 we have the following calculations:

$$P(E_1 \cup E_2 \mid F) = \frac{P((E_1 \cup E_2) \cap F)}{P(F)} = \frac{P(E_1 \cup E_2)}{P(F)}.$$

Due to the additivity of P we have:

$$\frac{P(E_1 \cup E_2)}{P(F)} = \frac{P(E_1) + P(E_2)}{P(F)} = P(E_1 \mid F) + P(E_2 \mid F).$$

The last property also follows easily from the conditional probability formula, and the fact that $P(E \cap F) = P(E)$ since $E \subseteq F$. $\qquad\square$

This theorem implies that $P(E \mid F)$ induces the probability space (Ω_F, P_F), where $\Omega_F = F$ is the new sample space and $P_F(E) = P(E \mid F)$ for all events $E \subseteq F$. This new probability space can be viewed as a cut-down version of the original probability space to the sub-sample space Ω_F. Note that for the events E that are subsets of F, they are more likely to occur in the new probability space (Ω_F, P_F) than in the old probability space (Ω, P). This is attested by the last part of the theorem above.

Note that from the conditional probability formula, for events E and F we can compute the event $P(E \cap F)$ by any of the following formulas:

$$P(E \cap F) = P(F) \cdot P(E \mid F) \quad \text{or} \quad P(E \cap F) = P(E) \cdot P(F \mid E).$$

Example 33.6. A box contains three red notebooks and five black notebooks. All are identical size wise. Two notebooks are taken one after the other at random with no replacement. Find the probabilities of the following events:

(1) Both notebooks are black.
(2) Both notebooks are red.
(3) One notebook is black and the other is red.

To solve this problem we need to define the sample space Ω and the probability P on it. We enumerate the notebooks as 1, 2, ..., 8 of which 1, 2 and 3 are red and the rest are black. Taking two books one after the other is thus a pair (i,j), where i and j are books and $i \neq j$. Hence, our sample space Ω is the following:

$$\Omega = \{(i,j) \mid 1 \leq i \leq 8 \text{ and } 1 \leq j \leq 8 \text{ and } i \neq j\}.$$

There are are 56 outcomes of the sample space. Hence the probability of each outcome is $1/56$. The event "both notebooks are black" corresponds to the set

$$E_1 = \{(i,j) \mid \text{both } i \text{ and } j \text{ are black}\}.$$

The event "both notebooks are red" corresponds to the set

$$E_2 = \{(i,j) \mid \text{both } i \text{ and } j \text{ are red}\}.$$

The event "one notebook is black and the other is red" corresponds to the set

$$E_3 = \{(i,j) \mid i \text{ and } j \text{ are of different color}\}.$$

It is easy to see to compute the number of outcomes in each of these events: $|E_1| = 20$, $|E_2| = 6$, $|E_3| = 30$. Hence, we now can easily compute the probabilities of these events:

$$P(E_1) = \frac{20}{56} = \frac{5}{14}, \ P(E_2) = \frac{6}{56} = \frac{3}{28}, \text{ and } P(E_3) = \frac{30}{56} = \frac{15}{28}.$$

This solves the posed problem.

We now recast the example above through conditional probability. For this, we define the following four events:

- $E_1^L = \{(i,j) \mid i \text{ is black}\}$ and $E_1^R = \{(i,j) \mid j \text{ is black}\}$.
- $E_2^L = \{(i,j) \mid i \text{ is red}\}$ and $E_2^R = \{(i,j) \mid j \text{ is red}\}$.

It is clear that the following equalities hold:

$$E_1 = E_1^L \cap E_1^R, \ E_2 = E_2^L \cap E_2^R, \text{ and } E_3 = E_1^L \cap E_2^R \cup E_1^R \cap E_2^L.$$

Now note the following probabilities:

$$P(E_1^L) = P(E_1^R) = \frac{5}{8}, \ P(E_2^L) = P(E_2^R) = \frac{3}{8},$$

and

$$P(E_1^L \cap E_2^R) = P(E_1^R \cap E_2^L) = \frac{15}{56}.$$

The probabilities of E_1 and E_2 can now be expressed through the conditional probability formula as follows:

$$P(E_1) = P(E_1^L) \cdot P(E_1^R \mid E_1^L) \text{ and } P(E_2) = P(E_2^L) \cdot P(E_2^R \mid E_2^L).$$

Now we can compute all the conditional probabilities:

$$P(E_1^R \mid E_1^L) = \frac{P(E_1^R \cap E_1^L)}{P(E_1^L)} = \frac{5/14}{5/8} = \frac{4}{7},$$

and

$$P(E_2^R \mid E_2^L) = \frac{P(E_2^R \cap E_2^L)}{P(E_2^L)} = \frac{3/28}{3/8} = \frac{2}{7}.$$

Thus,

$$P(E_1) = P(E_1^L) \cdot P(E_1^R \mid E_1^L) = \frac{5}{8} \cdot \frac{4}{7} = \frac{5}{14}$$

and

$$P(E_2) = P(E_2^L) \cdot P(E_2^R \mid E_2^L) = \frac{3}{8} \cdot \frac{2}{6} = \frac{3}{28}.$$

33.3   Independence

Let us consider Example 33.4 again. In this example, we flip a coin 4 times where the tail and head outcomes are equally likely to occur. We would like to know the probability that an even number of tails appear given that the second flip is tail. In this example, F is the event representing all outcomes in which the second flip is tail, and E is the event representing all outcomes with an even number of tails. We computed the conditional probability $P(E \mid F)$:

$$P(E \mid F) = \frac{P(E \cap F)}{P(F)} = \frac{0.25}{0.5} = \frac{1}{2}.$$

If we compute the probability of E in the whole sample space Ω, we clearly see that $P(E)$ remains equal to the condition probability $P(E \mid F)$, that is:

$$P(E) = P(E \mid F) = \frac{1}{2}.$$

In other words, the probability of the event E occurring does *not* depend on whether the event F occurred. This example shows that, in principle, it maybe the case that the equality $P(E \mid F) = P(E)$ holds for some events E and F in a given probability space. We single out such events in the next definition:

Definition 33.3. Let (Ω, P) be a probability space and $P(F) > 0$. We say that an event E **is independent of the event** F if $P(E \mid F) = P(E)$.

If E is independent of F, where $P(F) > 0$, then we have the following equalities:

$$P(E \mid F) = \frac{P(E \cap F)}{P(F)} = P(E), \text{ and, hence } P(E \cap F) = P(E) \cdot P(F).$$

Therefore if E is independent of F then F is independent of E. The independence of E and F is equivalent to saying that $P(E \cap F) = P(E) \cdot P(F)$.

Let us now consider Example 33.5. In this example let E be the event saying that the both children are girls. Recall that the sample space Ω in that example is $\{BB, BG, GB, GG\}$. Hence, the probability $P(E)$ of the event is $1/4$. However, under condition that one of the children is a girl implies that $P(E \mid F) = 1/3$. So, the event E is dependent of the event F.

33.4 Probability distribution

Assume that we have an experiment with outcomes $o_1, \ldots, o_n$. In order to introduce a probability into this sample space of outcomes, one needs to associate with each outcome its probability. For example, consider a throw of a six-faced fair die. The outcomes are $1, 2, \ldots, 6$. Assume that we declare that the probability of each outcome is $\frac{1}{6}$. This assignment allows us to define probabilities for the events of the sample space. We formalize this simple idea in the following definition.

Definition 33.4. Let $\Omega = \{o_1, \ldots, o_n\}$ be a sample space. A **probability distribution** for this sample space is any function μ that associates with each outcome o its probability $\mu(o) \geq 0$. Moreover, the function μ must satisfy the following equality:

$$\mu(o_1) + \mu(o_2) + \ldots + \mu(o_n) = 1.$$

The following is an easy proposition.

Proposition 33.1. *Every probability distribution μ determines the probability on the sample space Ω.*

Proof.
The proof is immediate. Indeed, for every event $E = \{e_1, \ldots, e_k\} \subseteq \Omega$ set the probability of E to be $\mu(e_1) + \ldots + \mu(e_k)$. Since this probability depends on the probability distribution μ, we denote the probability by P_μ. Thus,

$$P_\mu(E) = \mu(e_1) + \ldots + \mu(e_k).$$

Clearly, we now have the probability space (Ω, P_μ). □

If $\mu(o) = (1/n)$ for every $o \in \Omega$, where Ω has exactly n outcomes, then we say that μ is a **uniform probability distribution**. Thus, it is clear that uniform probability distributions represent those probabilities in which all outcomes are equally likely to occur.

Bernoulli distributions are well-known in probability theory and statistics. The distribution is named after Swiss scientist Jacob Bernoulli. We define Bernoulli distributions below:

Definition 33.5. Consider an experiment with exactly two outcomes, say 0 and 1, where 1 represents a success and 0 a failure. Any probability distribution of this sample space is called a **Bernoulli distribution**. Experiments with Bernoulli distributions are called **Bernoulli trials**.

Let μ be a Bernoulli distribution such that $\mu(1) = p$. Then $\mu(0) = q$ where $q = 1 - p$. Thus, the probability of a success in a trial is p, and the probability of a failure is q. When $p = 0.5$ the distribution clearly becomes uniform.

So, let μ be a Bernoulli distribution such that the probability of success is p, and failure is $q = 1 - p$. We now fix $n \geq 1$, and perform an experiment where we perform Bernoulli trials exactly n times. Let Ω_n be the set of outcomes of this experiment. The outcome $o \in \Omega_n$ is of the form $o = (a_1, \ldots, a_n)$ where each a_i is 0 or 1. Let i be the number of successes in o. Then $n - i$ is the number of failures. One can naturally associate the real number $\mu_n(o)$ to this outcome as follows:

$$\mu_n(o) = p^i \cdot (1 - p)^{n-i}.$$

For instance, if $\mu(0) = 1/2$ then $\mu(o) = 1/2^n$ or if $\mu(0) = 1/3$ the for $o = (1, 1, 1, 0, \ldots, o)$ we have $\mu_n(o) = (2/3)^3 \cdot (1/3)^{n-3}$. Our goal is to prove the following theorem.

Theorem 33.4. *The function μ_n is a probability distribution on the sample space Ω_n.*

Proof.
To prove the theorem, we use counting techniques from the previous lectures. Note that there are exactly $C(n, i)$ outcomes with i successes. Therefore, for the event $E_i = \{o \mid o \text{ has exactly } i \text{ successes}\}$, we have

$$P_\mu(E_i) = C(n, i) \cdot p^i \cdot (1 - p)^{n-i}.$$

Now using the formula

$$(x + y)^n = \sum_{i=0}^{n} C(n, i) x^i y^{n-i}$$

from Lecture 32 we have the following calculations:

$$\sum_{o \in \Omega_n} \mu(o) = \sum_{i=0}^{n} C(n, i) p^i \cdot (1 - p)^{n-i} = (p + (1 - p))^n = 1.$$

This shows that the function μ is a probability distribution on Ω_n. □

Example 33.7. Assume that we flip a biased coin 8 times where the probability of success is $2/3$. What is the probability of exactly 5 successes in this experiment?

There are $2^8 = 256$ possible outcomes in the sample space. We compute the probability distribution on this sample space. In an outcome o, if there are i successes and j failures then $j = 8 - i$ and the probability $\mu(o)$ of the outcome is then $\mu(o) = (2/3)^i \cdot (1/3)^{8-i}$. Hence, if there are exactly 5 successes then $\mu(o) = (2/3)^5 \cdot (1/3)^3$. The number of successful outcomes is exactly $C(8, 5)$. Therefore, for the event $E = \{o \mid o \text{ has exactly } 5 \text{ successes}\}$, we have

$$P_\mu(E) = C(8, 5) \cdot (2/3)^5 \cdot (1/3)^3 = 56 \cdot (2^5/3^8) = 7 \cdot (2^8/3^8) \approx 0.273.$$

33.5 Random variables

Consider an experiment when a pair of six-sided dice is rolled. Each out-come of the experiment can be presented as a pair (i, j), where the ranges of both i and j are $1, 2, 3, 4, 5, 6$. Suppose we are interested in the sum of the dice, and let $X(i, j)$ be the sum $i + j$ of the outcome (i, j). So, X takes the following values:

X(1,1)=2,
X(1,2)=X(2,1)=3,
X(1,3)=X(3,1)=X(2,2)=4,
X(1,4)=X(4,1)=X(2,3)=X(3,2)=5,
X(1,5)=X(5,1)=X(2,4)=X(4,2)=X(3,3)=6,
X(1,6)=X(6,1)=X(2,5)=X(5,2)=X(3,4)=X(4,3)=7,
X(2,6)=X(6,2)=X(3,5)=X(5,3)=X(4,4)=8,
X(3,6)=X(6,3)=X(4,5)=X(5,4)=9,
X(4,6)=X(6,4)=X(5,5)=10,
X(5,6)=X(6,5)=11,
X(6,6)=12.

Viewing X as a variable we can pose the following types of questions. What is the probability that the value of X is 2? What is the probability that the value of X is greater than 5? What is the probability that the value of X is strictly between 4 and 9? For example, the probability that the value of X is greater than 5, written $P(X > 5)$, is $\frac{26}{36} = \frac{13}{18}$. Similarly, the probability that the value of X is strictly between 4 and 9, written $P(4 < X < 9)$, equals to $20/36 = 5/9$.

We can also ask more sophisticated questions about the variable X. For example, what is the expected value of X? An answer to this question requires the definition of the expected value of X. Intuitively, the expected value of X is the value that we expect X to take, on average. If all the values of X occured with the same probability, then we would simply take the average of the values. However, we know from above that this is not the case. For example, $P(X = 2) = \frac{1}{36}$, whereas $P(X = 7) = \frac{1}{6}$. Therefore, we calculate the weighted average of the values, where the weight is the probability of that value occurring. In other words, for our example, we define the expected value of X, denoted by $E(X)$, to be the following sum:

$$E(X) = 2 \cdot P(X = 2) + 3 \cdot P(X = 3) + \ldots + 11 \cdot P(X = 11) + 12 \cdot P(X = 12).$$

We formalize this discussion as follows.

Definition 33.6. A **random variable** in a probability space (Ω, P) is a function that associates to each outcome in Ω a real number. We denote random variables by letters X, Y, etc.

For a random variable X and a number r, we can, for instance, consider the following probabilities:

$$P(X \leq r), \ P(X = r), \text{ and } P(X > r).$$

For example, $P(X \leq r)$ represents the probability of the following event $\{o \in \Omega \mid X(o) \leq r\}$.

The function X defined above is an example of a random variable. Here is another example.

Example 33.8. Suppose that a fair coin is flipped exactly n times. To each outcome $o = (a_1, \ldots, a_n)$ of this experiment, we associate $X(o)$ the number of 1's that occur in the outcome o. It is not hard to see, see for instance Example 33.3, that for each $i \leq n$ we have $P(X = i) = C(n, i)/2^n$.

Let X be a random variable in the probability space (Ω, P). We would like to compute the expected value of X when a large number of experiments are performed. The concept of the expected value of X is defined as follows.

Definition 33.7. The **expected value** of a random variable X, denoted by $E(X)$, on the probability space (Ω, P) is

$$E(X) = \sum_{i=1}^{n} x_i \cdot P(X = x_i)$$

where $x_1, \ldots,$ and x_n are all the values in the range of X. The term $x_i \cdot P(X = x_i)$ is called the *weighted value of x_i*. The weight here is $P(X = x_i)$.

Example 33.9. A fair coin is flipped three times. The sample space Ω consists of 8 outcomes such that the probability of each outcome is $1/8$. Let X be the random variable that computes the number of tails in outcomes. Find the expected value of X.

Let T and H denote the tail and the head of the coin, respectively. Note that $X(TTT) = 3$, $X(TTH) = X(THT) = X(HTT) = 2$, $X(THH) = X(HTH) = X(HHT) = 1$, and $X(HHH) = 0$. Therefore:

$$E(X) = 0 \cdot \frac{1}{8} + 1 \cdot \frac{3}{8} + 2 \cdot \frac{3}{8} + 3 \cdot \frac{1}{8} = \frac{3}{8} + \frac{6}{8} + \frac{3}{8} = \frac{12}{8} = \frac{3}{2}$$

Example 33.10. Consider Bernoulli trials where the probability of success is p (and remember that $q = 1 - p$). What is the expected number of successes when n Bernoulli trials are performed?

For this example, the probability space Ω consists of n-tuples $(a_1, \ldots, a_n)$, where each a_i is either 1 or 0. Our random variable X, given an outcome o, computes the number of successes in o. By, definition of $E(X)$ we have:

$$E(X) = \sum_{k=1}^{n} k \cdot P(X = k).$$

From the proof of Theorem 33.4 in previous section, we know that $P(X = k)$ is equal to $C(n, k) \cdot p^k \cdot (1 - p)^{n-k}$. Now note that $C(n, k) = nC(n - 1, k - 1)/k$. Indeed, this can easily be seen from the following calculation:

$$C(n, k) = \frac{n!}{k!(n - k)!} = \frac{(n - 1)!}{(k - 1)!((n - 1) - (k - 1))!} \cdot \frac{n}{k} = C(n-1, k-1) \cdot \frac{n}{k}.$$

Using these equalities, we compute $E(X)$:

$$E(X) = \sum_{k=1}^{n} k \cdot P(X = k) = n \sum_{k=1}^{n} C(n - 1, k - 1)p^k q^{n-k} = np \sum_{k=1}^{n} C(n - 1, k - 1)p^{k-1}q^{n-k}.$$

Now we simply adjust k by replacing it with variable j that ranges from 0 to $n - 1$, and use Theorem 32.4 that decomposes $(p + q)^m$:

$$E(X) = np \sum_{k=1}^{n} C(n-1, k-1)p^{k-1}q^{n-k} = np \sum_{j=0}^{n-1} C(n-1, j)p^j q^{n-1-j} = np(p + q)^{n-1} = np.$$

33.6 Exercises

(1) Prove that if $E \subseteq F$ in a probability space (Ω, P) then $P(E) \leq P(F)$.
(2) A box has three red and two black balls. A set of three balls is removed at random. Compute the probabilities for each of the following events:

   (a) all are red,
   (b) all are black,

(c) one is red and two are black, and

(d) one is black and two are red.

(3) A **palindrome** is a string w such that when we write w in reverse order the resulting string is again w. For example, 101, 1001, and 10100101 are palindromes. What is the probability that a given binary string of length n is a palindrome?

(4) Consider a poker hand (see Exercises of Lecture 32). Find probabilities of getting the following hands:

(a) four of a kind,

(b) two pairs,

(c) three of a kind, and

(d) one pair.

(5) Prove that for pairwise disjoint events $E_1, \ldots, E_k$ in probability space (Ω, P), we have:

$$P(E_1 \cup \ldots \cup E_k) = P(E_1) + \ldots + P(E_k).$$

(6) Prove that for events $E_1, \ldots, E_k$ in probability space (Ω, P), we have:

$$P(E_1 \cup \ldots \cup E_k) \leq P(E_1) + \ldots + P(E_k).$$

(7) Let (Ω, P) be a probability space.

(a) Prove that $P(E \cap F) \geq P(E) + P(F) - 1$.

(b) Extend the inequality above to

$$P(A_1 \cap \ldots \cap A_k) \geq P(A_1) + \ldots + P(A_k) - (k - 1).$$

(8) What is the conditional probability that a family with three children has exactly two boys given that one of the children is a boy?

(9) Consider the Example 33.6. Determine if the events E_1 and E_2 are independent of the event E_3.

(10) Let (Ω, P) be a probability space. Assume that $A_1, A_2, \ldots, A_k$ are pairwise disjoint events such that

(a) $\Omega = A_1 \cup A_2 \cup \ldots \cup A_k$, and

(b) $P(A_i) > 0$ for all $i = 1, \ldots, k$.

Prove the following equality:

$$P(B) = P(A_1) \cdot P(B \mid A_1) + P(A_2) \cdot P(B \mid A_2) + \ldots + P(A_k) \cdot P(B \mid A_k).$$

(11) Let $A_1, A_2, \ldots, A_k$ be events in a probability space. Prove the following equality:

$$P(A_1 \cap A_2 \cap \ldots \cap A_k) = P(A_1) \cdot P(A_2 \mid A_1) \cdot \cdots \cdot P(A_k \mid A_1 \cap \ldots \cap A_{k-1}).$$

(12) Consider Example 33.6. Determine if the events considered in this example are pairwise dependent or not.

Programming exercises

(1) Design classes for sample spaces, events, and random variables. Make sure to support:

(a) calculating the probability of an event,
(b) calculating the expected value of a random variable,
(c) computing the intersection, union, and negation of events, and
(d) calculating the conditional probability of an event given another event.

Solutions to selected exercises

> *Love is the answer to a question that I have forgotten.*
> Regina Spektor.

In this section, we present solutions to selected exercises from the lectures. Note that some of the answers here are intended to be guidelines, rather than full, detailed solutions.

Lecture 1

Question 2: (c) 2^n.

Question 2: (d) $2^{n+1} - 1$.

Question 4: (b) Let x and y be rational numbers. Then there exist integers n, m, s, t such that $x = \frac{m}{n}$, $y = \frac{s}{t}$ and $n \neq 0$, $t \neq 0$. So we can write

$$p - q = \frac{m \cdot t - s \cdot n}{n \cdot t}.$$

Since $m \cdot t - s \cdot n$ and $n \cdot t$ are integers and $n \cdot t \neq 0$, $p - q$ is a rational number by definition.

Question 4: (c) The statement is not a theorem. A counterexample is $x = 1$ and $y = 0$.

Lecture 2

Question 1: (f) Assume, by contradiction, that n is odd. Hence n is of the form $2k + 1$ for some integer k. Therefore $n^2 = 2(2k^2 + k) + 1$. Hence, n^2 is an odd number that contradicts the assumption.

Question 3: Assume that $r = \sqrt{n^2 + m^2}$ is a rational number. Then r is of the form s/t, where s and t are integers with $t \neq 0$. We can assume that t and s have no factors > 1 in common. Since n and m are odd we can write them as $n = 2a+1$ and $m = 2b+1$, where a and b are integers. Now, we do the following calculations:

$$s^2 = 2(2a^2 + 2b^2 + 2a + 2b + 1)t^2.$$

Hence s is even and can be written in the form $2k$. Therefore $s^2 = 4k^2$. Substituting this into the equality above gives:

$$2k^2 = (2a^2 + 2b^2 + 2a + 2b + 1)t^2.$$

The sum $2a^2 + 2b^2 + 2a + 2b + 1$ is odd. Hence t is even. But s and t were chosen not to have common factors great than 1. This is a contradiction.

Question 4: Consider $a = \sqrt{2}$ and $b = \sqrt{2}$. If a^b is a rational number then we have what is required. If a^b is irrational number then set $a = \sqrt{2}^{\sqrt{2}}$ abd $b = \sqrt{2}$. In this case, a^b equals 2. Again, we found two irrational numbers a and b such that a^b is a rational number.

Lecture 3

Question 2: Let $n = m \cdot k$ where $m > 1$ and is the smallest factor of n. If m is not prime then $m = p \cdot q$ where both p and q are positive and strictly less than m. In this case $p < m$ and p is a factor of n. But m was chosen to be the least factor. Contradiction.

Question 6: There are finitely many factors of a and of b. Among finitely many numbers there is always the greatest one.

Question 9: (a) Let $n = p_0 \cdot \ldots \cdot p_n + 1$. For each prime number p_i, where $i = 0, \ldots, n$, we can write n as $n = q_i \cdot p_i + 1$. Hence, none of these prime numbers divides n. Therefore n can not be written as a product of these prime numbers. But by the fundamental theorem of arithmetic n must be divided by a prime number, call it q. This q is distinct from all the prime numbers $p_0, p_1, \ldots, p_n$.

Lecture 4

Question 4: Congruence classes modulo 5 are:

$\bar{0} = \{\ldots, -15, -10, -5, 0, 5, 10, 15, \ldots\},$
$\bar{1} = \{\ldots, -14, -9, -4, 1, 6, 11, 16, \ldots\},$
$\bar{2} = \{\ldots, -13, -8, -3, 2, 7, 12, 17, \ldots\},$
$\bar{3} = \{\ldots, -12, -7, -2, 3, 8, 13, 18, \ldots\},$
$\bar{4} = \{\ldots, -11, -6, -1, 4, 9, 14, 19, \ldots\}.$

Lecture 5

Question 1: We know that n is congruent to n' and m is congruent to m' modulo p. This means that there are integers k and l such that $n - n' = kp$ and $m - m' = lp$. We can write that $n = n' + kp$ and $m = m' + lp$. Now we have that

$$n \cdot m - n' \cdot m' = (n'+kp)(m'+lp) - n' \cdot m' = n' \cdot m' + n' \cdot lp + m' \cdot kp + p^2 kl - n' \cdot m' = p(n' \cdot l + m' \cdot k + p \cdot kl).$$

As one can see p divides $n \cdot m - n' \cdot m'$. Therefore $n \cdot m$ is congruent to $n' \cdot m'$ modulo p.

Question 2: For $+$ modulo 5:

$+$	$\bar{0}$	$\bar{1}$	$\bar{2}$	$\bar{3}$	$\bar{4}$
$\bar{0}$	$\bar{0}$	$\bar{1}$	$\bar{2}$	$\bar{3}$	$\bar{4}$
$\bar{1}$	$\bar{1}$	$\bar{2}$	$\bar{3}$	$\bar{4}$	$\bar{0}$
$\bar{2}$	$\bar{2}$	$\bar{3}$	$\bar{4}$	$\bar{0}$	$\bar{1}$
$\bar{3}$	$\bar{3}$	$\bar{4}$	$\bar{0}$	$\bar{1}$	$\bar{2}$
$\bar{4}$	$\bar{4}$	$\bar{0}$	$\bar{1}$	$\bar{2}$	$\bar{3}$

Lecture 6

Question 3:
(a) At least the node v itself is contained in the component, therefore it is not empty and contains v (the fact that the component exists is straightforward from its definition).

(b) Assume there is w which belongs to both $C(v)$ and $C(u)$. Then every x in $C(u)$ is strongly connected to w and thus to v. This implies that all vertices in $C(u)$ are in $C(v)$. Similarly, all vertices in $C(v)$ are in $C(u)$. Hence, the components $C(u)$ and $C(v)$ are the same components.

Lecture 7

Question 3: Let $n = 2k > 2$ be an even number. Then G_n, a 3-regular graph with n vertices, can be constructed as follows. Let $1, 2, \ldots, 2k$ be the vertices of G_n. Now, for every $1 \leqslant i \leqslant 2k - 1$, add an edge $\{i, i+1\}$ to the graph. Also add an edge $\{1, 2k\}$. Then for each $1 \leqslant i \leqslant k$ add an edge $\{i, i+k\}$ to G_n.

Question 4: For the first part, each vertex v on its own forms a path of length 0. So, v is connected to itself. For the second part, since v is connected to u, there is a path $v_0, \ldots, v_n$ from v to u. Then $v_n, \ldots, v_0$ will be a path from u to v. Hence, u is connected to v. For the last part, since v is connected to u, there is a path $v_0, \ldots, v_n$ from v to u. Since u is connected to w then there is a path $u_0, \ldots, u_m$ from u to w. Then $v_0, \ldots, v_n, u_0, \ldots, u_m$ will be a path from v to w. Hence, v is connected to w.

Question 5: The degree of a vertex v is equal to the number of edges adjacent to it. Consider any edge, say $\{a, b\}$. This edge is adjacent to a and b. So, when we compute $\sum_{v \in V} \deg(v)$, we count this edge twice. Once when we compute $\deg(a)$ and the next time when we compute $\deg(b)$. Therefore, the sum $\sum_{v \in V} \deg(v)$ is equal to 2 times the number of edges in G, which is of course an even number.

Lecture 8

Question 5: For part (a), consider K_1, it consists only of one vertex v and v is a Hamiltonian circuit of length 0. Now, consider K_2. It does not have a Hamiltonian circuit. Now, let n be greater then 2. By definition of K_n, any sequence of all vertices $v_1, \ldots, v_n, v_1$, where all vertices $v_1, \ldots, v_n$ are pairwise distinct, forms a Hamiltonian circuit. For part (b), the degree of every vertex in K_n is equal to $n - 1$. Therefore, when n is odd K_n has an Euler circuit. When n is even K_n does not have an Euler curcuit.

Lecture 9

Question 4: The proof is a word by word repetition of the second and third paragraphs of the proof of Proposition 9.1 replacing $+$ with $-$ and $*$, respectively.

Question 8: An example of a finitely branching infinite tree is the set N of natural numbers in which the root is 0, and the child of node i is $i + 1$.

Question 9: Let T be an infinite finitely branching tree. Construct the following path by induction.

Base case. The first element x_0 of the path is the root of the tree. There are infinitely many nodes below x_0.

Inductive hypothesis. Assume we have constructed a path $x_0, x_1, \ldots, x_n$. The inductive hypothesis on this path is that there are infinitely many nodes below x_n.

Inductive step of the construction of the path. Consider all the children y_1, y_2, ..., y_s of x_n. Clearly, s is a natural number greater than 0. It must be the case that one of the children has infinitely many nodes below it. Indeed, if every child y_i of x_n had finitely many nodes below it, then there would be finitely many nodes below x_n, which contradicts the inductive hypothesis. So, take a child y of x_n such that y has infinitely many nodes below it. Set $x_{n+1} = y$. We have constructed a path $x_0, x_1, \ldots, x_n, x_{n+1}$ such that there are infinitely many nodes below x_n.

Consider the sequence $x_0, x_1, x_2, x_3, \ldots, x_n, x_{n+1}, \ldots$. This is an infinite sequence that is also a path. This shows that every infinite finitely branching tree has an infinite path.

Lecture 10

Question 3: **b)** $(A \cap B) \cap C = A \cap (B \cap C)$. At first, show that $(A \cap B) \cap C \subseteq A \cap (B \cap C)$. Let $x \in (A \cap B) \cap C$. Then $x \in (A \cap B)$ and $x \in C$. Therefore, $x \in A$ and $x \in B$ and $x \in C$. Hence, $x \in A$ and $x \in (B \cap C)$. Therefore, $x \in A \cap (B \cap C)$. The proof of the inclusion $(A \cap B) \cap C \supseteq A \cap (B \cap C)$ is similar.

Question 5: **a)** Prove that $P(A \cap B) = P(A) \cap P(B)$. Let $C \in P(A \cap B)$. By definition, $C \subseteq A \cap B$. Hence, $C \subseteq A$ and $C \subseteq B$. Again, by definition, $C \in P(A)$ and $C \in P(B)$. Therefore, $C \in P(A) \cap P(B)$. We have proved that $P(A \cap B) \subseteq P(A) \cap P(B)$. Now, prove that $P(A \cap B) \supseteq P(A) \cap P(B)$. Let $C \in P(A) \cap P(B)$. Then $C \in P(A)$ and $C \in P(B)$. By definition, $C \subseteq A$ and $C \subseteq B$. Therefore, $C \subseteq A \cap B$ and $C \in P(A \cap B)$.

b) We disprove that $P(A \cup B) = P(A) \cup P(B)$. Let $A = \{1, 2\}$, $B = \{1, 3\}$ and $C = \{1, 2, 3\}$. As one can see $C \subseteq A \cup B$. Hence, $C \in P(A \cup B)$. Now, show that $C \notin P(A) \cup P(B)$. Suppose otherwise, then $C \in P(A)$ or $C \in P(B)$. Therefore, $C \subseteq A$ or $C \subseteq B$, which is impossible. This contradiction shows that $C \notin P(A) \cup P(B)$. Therefore, $P(A \cup B) \neq P(A) \cup P(B)$.

Lecture 11

Question 3: The number of ternary relations on $\{x, y\}$ is 2^8, and the number of relations of arity 4 is 2^{16}.

Question 4: The relations S is not reflexive, not symmetric, not transitive, and not antisymmetric. The relation T is reflexive, symmetric, antisymmetric, and transitive.

Question 11: For any element $X \in P(A)$, we have $X \sim X$ because $X \setminus X$ has 0 elements. So, $\sim$ is reflexive. For $X, Y \in P(A)$ if $X \sim Y$ then $X \setminus Y$ and $Y \setminus X$ are both finite. Then clearly, $Y \setminus X$ and $X \setminus Y$ are both finite. Hence $Y \sim X$. This shows that $\sim$ is symmetric. To show transitivity of $\sim$, assume that for $X, Y, Z \in P(A)$ we have $X \sim Y$ and $Y \sim Z$. Then

$$X \setminus Z \subseteq X \setminus Y \cup Y \setminus Z.$$

Hence $X \setminus Z$ is finite. Similarly, $Z \setminus X$ is finite. Therefore, $X \sim Z$. This shows transitivity of $\sim$.

Lecture 12

Question 1: (b) The union of equivalence relations is not necessarily an equivalence relation. For instance, consider $E_1 = \{(a,a), (b,b), (c,c), (a,c), (c,a)\}$ and $E_2 = \{(a,a), (b,b), (c,c), (a,b), (b,a)\}$. Both are equivalence relations on $\{a, b, c\}$ but $E_1 \cup E_2$ is not an equivalence relation.

Question 2: There are 19 partial orders on the set $\{a, b, c\}$. (To see them just draw their Hasse diagrams).

Question 9: The relation R is reflexive because $x = 2^0 x$. Hence $(x, x) \in R$. Assume that $(x, y) \in R$ and $(y, x) \in R$. Then $y = 2^n x$ and $x = 2^m y$ for some $n, m \in \mathbb{N}$. So, we have $y = 2^n x = 2^n 2^m y = 2^{n+m} y$. This can only happen when either $x = y = 0$ or $n = m = 0$. Hence, R is anti-symmetric. Assume that $(x, y) \in R$ and $(y, z) \in R$. Then $2^n x = y$ and $2^m y = z$ for some $n, m \in \mathbb{N}$. Hence $z = 2^{n+m} x$. Therefore R is transitive.

The minimal elements of the partial order are all odd numbers.

Lecture 13

Question 2: One needs to remove the edges in the graph representation G of R, and add edges (x, y) if there were no edge from x to y in G.

Question 5: (b) The method needs to simply add a record (x, y) into the table whenever (x, y) belongs to both R_1 and R_2.

Lecture 14

Question 2: Take any k-tuple $(x_1, \ldots, x_k)$ over the underlying domain A of R. If $(x_1, \ldots, x_k) \in R$ then $(x_1, \ldots, x_k, x) \in c(R)$ for any

$x \in A$. Hence, by the definition of $c(R)$, we have $(x_1, \ldots, x_k) \in \exists x_{k+1} c(R)$. If $(x_1, \ldots, x_k) \in \exists x_{k+1} c(R)$ then there must exists an $x \in A$ such that $(x_1, \ldots, x_k, x) \in c(R)$. This can only happen when $(x_1, \ldots, x_k) \in R$.

Lecture 15

Question 2: One possible way to prove correctness of the $PathExistence(G, s, t)$ algorithm is to ensure that the following statement is a loop invariant. *A vertex is marked visited at iteration t if and only if there exists a path of length at most t from s to the vertex.*

Question 3: *Hint.* Your statement should connect common divisors of n and m with common divisors of the values of the variables a and b.

Question 5: The total number of swaps is bounded by $n \cdot (n+1)/2$, where n is the size of the array **A**. Thus, the running time of the algorithm is bounded by a quadratic polynomial $p(n)$ (for input arrays of size n and p does not depend on n but the algorithm). If the running time of an algorithm is bounded by a polynomial, then such algorithms are theoretically considered to be practical.

Lecture 16

Question 4: For $n = 0$, in the base case, we have $11^0 - 4^0 = 7$. Hence the statement is true for the base case. Our inductive assumption is this. We assume that we have proven the statement for all $k < n$. Now our inductive steps consist of proving that the statement is true for the case when $n = k$.

Consider $11^n - 4^n$. We can write the following equalities:

$$11^n - 4^n = 11 \cdot (11^{n-1} - 4^{n-1}) + 11 \cdot 4^{n-1} - 4^n = 11(11^{n-1} - 4^{n-1}) + 4^{n-1}(11-4).$$

Each of the last two summands is divisible by 7. Hence $11^n - 4^n$ is also divisible by 7. Thus, by the first inductive principle, $11^n - 4^n$ is always divisible by 7 for all natural numbers n.

Question 7: For $n = 0$, in the base case, we have $0 = 2 \cdot 0 + 3 \cdot 0$. Hence the statement is true for the base case. Our inductive assumption is this. We assume that we have proven the statement for all $k < n$. Now our inductive steps consist of proving that the statement is true for the case when $n = k$.

Consider n. We write n as $(n-1)+1$. For $n-1$ we know that we already have integers a_1 and b_1 such that $n-1 = 2 \cdot a_1 + 3 \cdot b_1$. We can write 1 as $2 \cdot (-1) + 3 \cdot 1$. Therefore

$$n = (n-1)+1 = (2 \cdot a_1 + 3 \cdot b_1) + (2 \cdot (-1) + 3 \cdot 1) = 2 \cdot (a_1 - 1) + 3 \cdot (b_1 + 1).$$

Thus, by the first inductive principle, each n can always be written as $2 \cdot a + 3 \cdot b$ for some integers a and b.

Question 12: *Hint.* See answer to Question 5 for Lecture 15. The running time of the algorithm is bounded by roughly $n \cdot log(n)$ function for inputs of size n (that is, arrays with n items).

Lecture 17

Question 2: As an example we prove that (a) implies (b). We prove this by contradiction. Indeed, assume an edge e connecting x and y is removed from G but the remaining G is still connected. Then there exists a path $v_1, \ldots, v_n$, where $v_1 = x$ and $v_n = y$, from x to y that does not use the edge e. Then the path $v_1, \ldots, v_n, v_1$ is a cycle in the original graph G. Hence G is not a tree which contradicts the assumption.

Question 3: Let T' be the graph obtained from the tree T by removing the edge e. Assume that the edge e connects the nodes x and y. Consider the set T_1 of all the nodes that are connected to x in T'. Similarly, consider the set T_2 of all nodes that are connected to y in T'. Both T_1 and T_2 are non-empty sets. Moreover, every node is connected to either x or y in T'. Since x and y are not connected in T', the sets T_1 and T_2 have no nodes in common. Now, since T_1 and T_2 are connected and edges in both are edges of T, both T_1 and T_2 are actually trees.

Lecture 19

Question 5: The sets of winning vertices for the players are defined by induction. We outline the induction process that defines the winning set for Player 0, and leave the reader to check the correctness of the process.

Basis. Set $X_0 = T$.

Inductive Step. Our hypothesis is that we have defined the sets X_0, ..., X_n such that $X_0 \subseteq X_1 \subseteq \ldots \subseteq X_n$ and Player 0 wins from each of the vertices in X_n. For each vertex $v \notin X_n$ proceed as follows.

Case 1. If $v \in V_1$, put v into X_{n+1} if $v \notin A$, and for all u if $(v, u) \in E$ then $u \in X_n$.

Case 2. If $v \in V_0$, put v into X_{n+1} if $v \notin A$ and there exists a u such that $(v, u) \in E$ and $u \in X_n$.

Finally, put all vertices in X_n into X_{n+1}. If $X_{n+1} = X_n$ then stop the process.

Lecture 20

Question 3:

(a) $f \circ g(q) = 2(2q^2 - 1)/(2q^2 - 1)^2 + 1$.
(b) $g \circ f(q) = -(q^4 + 1)/2(q^2 + 1)^2$.

Lecture 21

Question 2: Assume that $f(x_1, y_1) = f(x_2, y_2) = (a, b)$. If both x_1 and y_1 both have the same parity then a and b have different parities. If x_1 and y_1 have different parities then a and b have the same parity. This implies that x_2 and y_2 have the same parity.

Assume that x_1 and y_1 are both odd. Then $b = y_1 + 1$ is even. Since $f(x_1, y_1) = f(x_2, y_2)$, we have x_2 and y_2 are both odd. Thus, we have $(x_1, y_1 + 1) = (x_2, y_2 + 1)$. Therefore $x_1 = x_2$ and $y_1 = y_2$. Similarly, $x_1 = x_2$ and $y_1 = y_2$ if both x_1 and y_1 are even.

Assume that x_1 is even but y_1 is odd. Then $a = x_1 + 1$ is odd. Since $f(x_1, y_1) = f(x_2, y_2)$ we have that x_2 is even and y_1 is odd. Therefore $x_1 = x_2$ and $y_1 = y_2$. Similary, $x_1 = x_2$ and $y_1 = y_2$ if x_1 is odd and y_1 is even.

Question 3: As an example, we prove that if A and B have the same number of elements then there is a bijection from A to B. Let us list all elements of A and B: $A = \{a_1, \ldots, a_n\}$ and $B = \{b_1, \ldots, b_m\}$. By assumption $n = m$. Define the following mapping $f : A \to B$: $f(a_i) = b_i$ for $i = 1, \ldots, n$. The mapping f is a bijection.

Question 5: Let S be a subset of A (that is, an element of $P(A)$). Using the subset S we define the following function $ch_S : A \to \{0, 1\}$. For all $x \in S$ set $ch_S(x) = 1$, and for all $x \in A$ but not in S, set $ch_S(x) = 0$. The mapping $S \to ch_S$ establishes the function from $P(A)$ into 2^S. It is left to the reader to show that this naturally defined function establishes a bijection between $P(A)$ and 2^A. The function ch_S is often called the **characteristic function** of the set S.

Question 11: The desired function is defined as follows. Consider an E-equivalence class $[n]$ of some number $n \in \mathbb{N}$. Map $[n]$ to the minimal number in the equivalence class $[n]$. Thus, we have a function

$$f : \mathbb{N}/E \to \{x \mid x \text{ is a representative of some } E\text{-equivalence class}\}.$$

The function f is onto because if x is a representative then $f([x]) = x$. If $[n]$ and $[m]$ are distinct E-equivalence classes then $min\ [n] \neq min\ [m]$. Hence,

$$f([n]) \neq f([m]).$$

Thus, f is a bijection.

Lecture 22

Question 2: We describe our algorithm without being too formal. Let s be a string over the alphabet $\{p_1, \ldots, p_n,), (, \vee, \wedge, \neq\}$. We call formulas of the type $\neg(p)$, $(p \vee q)$, $(p \wedge q)$, and $(p \to q)$ **atomic**, where $p, q \in PROP$. Thus, given the string s to detect if s is a formula, we proceed as follows:

(1) If $s \in PROP$, then declare that s is a formula.
(2) Find the first occurrence in s that contains an atomic sub-formula. If such an occurrence does not exist, then declare that s is not a formula.
(3) Replace the first occurrence of an atomic formula in s by a new proposition p.
(4) Set s' be the string obtained from the previous step. Set s to be s'.
(5) Repeat the process.

Note that the length of s is reduced at line 4. Also note that s is a formula if and only if s' is a formula. Hence, the algorithm detects if s is a formula.

Question 3: The proof goes by induction on the length of the formula ϕ. In the basis case, when the length of ϕ is 1 there is nothing to prove since formulas of length one are propositions. Suppose that we have proven the statement for all formulas of length strictly less than n.

Let ϕ be a formula of length $n > 1$. We now use the unique readability theorem. By the theorem ϕ can be uniquely written in one of the following forms:

(1) $\neg(\alpha)$.
(2) $(\alpha \lor \beta)$.
(3) $(\alpha \land \beta)$.
(4) $(\alpha \to \beta)$

In the first case, if (is the first occurrence in ϕ then the formula uniquely associated with it is α. Otherwise, (occurs in α itself, and hence the inductive assumption can now be applied.

In the second case, assume that (is the first occurence in ϕ. Then the formula ϕ itself is the one associated with (. Otherwise (occurs either in α or in β. In each case the inductive hypothesis is again applied.

Lecture 23

Question 1: There are exactly 8 truth assignments of the propositions p_1, p_2 and p_3. Consider, for instance, the following assignment A: $A(p_1) =$ **true**, $A(p_2) =$ **false** and $A(p_3) =$ **true**. Under this assignment, one can compute that $V_A(\phi) =$ **true**.

Question 2: This can be proved by induction on n. When we have one proposition, then there are exactly 2 truth assignments. Assume that there are 2^{n-1} assignments of $p_1, \ldots, p_{n-1}$. Each of these assignments A can be *extended* to assignment A_1 and A_2 as follows. Both A_1 and A_2 give the same value as A to propositions $p_1, \ldots, p_{n-1}$. In addition $A_1(p_n) =$ **true** and $A_2(p_n) =$ **false**. Hence, there are exactly 2^n assignments of $p_1, \ldots, p_n$.

Lecture 24

Question 1: 2^{2^n}.

Question 2: (b) $(p \land s \land \neg(q)) \lor (p \land q \land s)$.

Question 4: We prove part (2) of the proposition. If ϕ is satisfiable then there is a truth valuation A such that $V_A(\phi) = \{$**true**$\}$. This implies that ϕ can not be a contradiction. If ϕ is not a contradiction then there exists an assignment A such that does not evaluate ϕ to false. Hence A witnesses that ϕ is not a contradiction.

Question 6: We prove the induction case when ϕ is $(\phi_1 \land \phi_2)$. Assume that X is a model of ϕ. Then X is a model of ϕ if and only if X is a model of ϕ_1 and of ϕ_2. By inductive assumption, this is equivalent to saying that

A_X evaluates ϕ to **true** if and only if A_X evaluates both both ϕ_1 and ϕ_2 to **true**. Hence A_X evaluates ϕ to true.

Lecture 25

Question 2: (a). We prove that the *SubString* relation is reflexive, antisymmetric and transitive. For each string v, we know that v is a substring of v. Hence, we have reflexivity. For all strings u and v, assume that u is a substring of v and v is a substring of u. This implies that u and v have the same length. Since each is a substring of the other, this implies that $u = v$. Assume that u is a substring of v and v is a substring of w. This implies that we can write v as $v = x_1 u x_2$, and we can write w as $w = y_1 v y_2$. Hence $w = y_1 x_1 u x_2 y_2$. Thus, u is a substring of w.

Question 4: Examples of conditions that guarantee that $U^\star \cup V^\star = (U \cup V)^\star$ are any of the following.

(1) $U \subseteq V$ or $V \subseteq U$.
(2) $U^\star \subseteq V^\star$ or $V^\star \subseteq U^\star$.
(3) Each string in U can be written as a product of strings in V.
(4) Each string in V can be written as a product of strings in U.
(5) For all $u \in U$ and $v \in V$ the product uv is a product of strings from either U or V.

Question 5: The proof is by induction on the length of input strings. Indeed, let w be an input string of the form $\sigma_1 \ldots \sigma_n$, where $\sigma_i \in \Sigma$. If $n = 1$, then the run of $\mathcal{M}$ on w is simply the sequence $q_0, T(q_0, \sigma_1)$. Assume that for all strings v of length at most $k < n$, $\mathcal{M}$ has a unique run on v. Then for the string w, we write it as

$$\sigma_1 \sigma_2 \ldots \sigma_{n-1} \cdot \sigma_n.$$

Let $s_1, \ldots, s_n$ be the unique run of $\mathcal{M}$ on $\sigma_1 \ldots \sigma_{n-1}$. Then the following is the unique run of $\mathcal{M}$ on w:

$$s_1, \ldots, s_n, T(s_n, \sigma_n).$$

Question 6: There are many ways to prove that every finite language is FA recognizable. One way follows directly from the lecture where we construct a DFA recognizing the language $\{w\}$, where w is a finite string. We outline the construction and leave the details to the reader. So now assume that our finite language is the following:

$$\{w_1, \ldots, w_n\},$$

where $w_i = \sigma_{i,1} \ldots, \sigma_{i,k_i}$ with $i = 1, \ldots, n$. The states of the DFA we want to build are of the form:

$$(x_1, \ldots, x_n),$$

where $x_i = 0, 1, \ldots, k_i$. There is also one state **fail** which is a failing state. The initial state is $(0, \ldots, 0)$. The intuition here is that the ith component of the states is responsible for verifying if the string read equals w_i. If the ith component recognizes that the string read is w_i then the automaton goes into accepting state.

Lecture 26

Question 2: For $n = 1$, the language L_1 can be selected to be $\{0,1\}^\star$. For $n = 2$, the language L_2 can be the language $\{0,1\}^\star \setminus \{\lambda\}$. For $n > 2$, the language L_n can be selected to be $L_n = \{0^{n-2}\}$. It is clear that each L_n can be accepted by a DFA with n states. The rest is left to the reader.

Question 5: We prove part (a). Consider an ultimately periodic language $L = \{a^n, a^{n+p}, a^{n+2p}, \ldots, a^{n+ip}, \ldots\}$. A DFA accepting L is the following:

- The states are $0, \ldots, n, n+1, \ldots, n+(p-1)$.
- 0 is the initial state.
- n is the accepting state.
- Transitions are from state i to state $i+1$ for $i < n+(p-1)$, and then from state $n+(p-1)$ to n.

This DFA clearly accepts L.

Lecture 27

Question 5: Let $\mathcal{M} = (S, I, T, F)$ be an NFA accepting W. An automaton $\mathcal{M}' = (S', I', T', F')$ accepting $Prefix(W)$ is built as follows:

- The set S' of states is S.
- The set I' of initial states also remains unchanged.
- The transition diagram T' is also T.
- Declare a state s accepting if there exists a path from s to an accepting state in F in the automaton $\mathcal{M}$.

The reader is left to prove that the construction given is correct.

Lecture 28

Question 2: *Hint*: The proof that the DFA constructed is equivalent to the given NFA is the same as the proof of Theorem 28.1. The automaton constructed can be obtained from the automaton $\mathcal{B}$ from the proof of Theorem 28.1 by removing unreachable states from $\mathcal{B}$.

Lecture 29

Question 2: If $L = \{w_1, \ldots, w_n\}$ is a finite language, then the regular expression for it is $(w_1 + \ldots + w_n)$.

Lecture 31

Question 2: Let $A_1, \ldots, A_n$ be sets with $k_1, \ldots, k_n$ elements, respectively. We want to show that the set $B = A_1 \times \ldots \times A_n$ has $k_1 \cdot \ldots \cdot k_n$ elements. For $n = 1$, it is clear that B has k_1 elements. Similarly, for $n = 2$, the set $A \times A_2$ has $k_1 \cdot k_2$ elements. Assume that we have proved the statement for all sets $A_1, \ldots, A_t$, where $t < n$. Now consider the sets $A_1, \ldots, A_n$. The set $B' = A_1 \times \ldots \times A_{n-1}$ has $k_1 \cdot \ldots k_{n-1}$ elements by inductive assumption. Now, note that the set $B = A_1 \times \ldots \times A_n$ can be identified with the set $B' \times A_n$. Again, we apply our base case ($n = 2$) to B' and A_n and see that B has $k_1 \cdot \ldots \cdot k_n$ elements.

Question 4: Divide each $t + i$ by n and consider the remainder r_i. Since there are exactly n remainders $0, 1, \ldots, n - 1$, each member in the sequence $t + 1, t + 2, \ldots, t + p$ is obtained from the previous one by adding 1, and there are $p > n$ numbers in this sequence, it must be the case that one of the members of the sequence is divisible by n.

Question 5: The number of students should be greater than 18.

Question 7: One way to prove this is by induction. In the base case, when $n = 3$, an easy check shows that at least two people in the group of three people know the same number of people. Assume that the statement is true for any group of k people, where $k < n$. Consider now a group of n people $p_1, \ldots, p_n$. There are two cases. In case 1, we assume that in the group each person knows at least someone. Then with each person p_i we associate the number a_i that is equal to the number of people known by p_i. Clearly, $1 \le a_i \le n - 1$. By the pigeonhole principle, there are two people p_i and p_j such that $a_i = a_j$. In case 2, we assume that there is a person who knows no-one in the group. Say the person is p_1. Then no-one in the group knows p_1. Thus, we now can consider the group $p_2, \ldots, p_n$ of $n-1$ people. Now, the induction hypothesis can be applied.

Lecture 32

Question 2: The permutation Q is defined using P as follows: $Q(i) = j$ if and only if $P(j) = i$. The reader can check that $P \circ Q = Q \circ P = I_A$.

Question 4: The equality can be prove through a direct calculation as follows. We note that

$$C(n,k) = \frac{n!}{(k-1)! \cdot (n-k+1)!} \cdot \frac{n-k+1}{k} = C(n, k-1) \cdot \frac{n-k+1}{k}.$$

Similarly,

$$C(n,k) = \frac{n!}{k! \cdot (n-k)!} = \frac{(n-1)!}{k! \cdot (n-1-k)!} \cdot \frac{n}{n-k} = C(n-1, k)\frac{n}{n-k}.$$

Using these two equalities, we get the following:

$$C(n,k) = C(n, k-1) \cdot \frac{n-k+1}{k} = C(n-1, k-1) \cdot \frac{n}{n-k+1} \cdot \frac{n-k+1}{k}.$$

Thus, $C(n,k) = nC(n-1, k-1)/k$ which was required.

Question 5: 2^n equals the number of subsets of an n element set A. In turn $C(n,r)$ represents the number of subsets of A of size r. Therefore, the sum of all $C(n,r)$ is equal to the number of subsets of A.

Lecture 33

Question 1: If $E \subseteq F$ then we can write $F = E \cup (F \backslash E)$. These two sets in the union are pairwise disjoint. Hence $P(F) = P(E) + P(F \backslash E) \geq P(E)$.

Question 3: If n is even, there are exactly $2^{n/2}$ palindromes of length n. Hence the probability P that a string of length n is a palindrome is computed as follows:

$$P = 2^{n/2} \cdot \frac{1}{2^n} = \frac{1}{2^{n/2}}$$

If n is odd then there are exactly $2^{(n+1)/2}$ palindromes of length n. Hence, the probability P that a string of length n is a palindrome is computed as:

$$P = 2^{(n+1)/2} \cdot \frac{1}{2^n} = \frac{1}{2^{(n-1)/2}}$$

Question 7: (a) By Theorem 33.2 (b) and the fact that $1 \geq P(A)$ for all A, we have

$$1 \geq P(E \cup F) = P(E) + P(F) - P(E \cap F).$$

This inequality implies

$$P(E \cap F) \geq P(E) + P(F) - 1.$$